JACK APPRECIATES MATH

Revised First Edition

By Tuesday J. Johnson

University of Texas - El Paso

cognella®
academic publishing

Bassim Hamadeh, CEO and Publisher
Michael Simpson, Vice President of Acquisitions
Jamie Giganti, Senior Managing Editor
Miguel Macias, Graphic Designer
Marissa Applegate, Senior Field Acquisitions Editor
Gem Rabanera Project Editor
Elizabeth Rowe, Licensing Coordinator
Sean Adams, Production Editor

Printed in the United States of America

ISBN: 978-1-63487-622-3 (pbk)/ 978-1-63487-623-0 (br)

www.cognella.com   800-200-3908

CONTENTS

PREFACE

How to use this text as a student: This is not a standard mathematics text: it is meant to be read. I know, and all other math professors and teachers know, that all mathematics texts are meant to be read; however, this one is written in story form. The exercises are presented at the beginning of the section, laying out the story of a guy named Jack with all the math he encounters throughout his life. After the exercises comes the mathematics that will help you help Jack. I have found that most of my students go to the exercises first, then go back and look for examples if they do not understand the instructions or skills involved. This book is written in that more frequently used order. Each grouping of exercises will have one or more sections of the lesson that help you to solve the problems. The most important thing you can do as a student using this book is to read. Read the exercises first, read the lessons next, finally watch the suggested videos and read any supplemental information you can, as necessary. As you read the text portion, work out the examples as they are presented before looking at the solutions provided. This is your chance to practice the skills.

How to use this text as a professor: I generally break the chapters into groups based on the exercises. Suggested assignments have been marked by stick figures of the main character. The same image can be found in the reading to link the materials together. The amount of material you cover each day depends on your preferences and student abilities. I assign the reading of the lessons prior to each class lecture along with a summary due before class time. The summary requires the students to tell me what they already knew, what they learned, and what they did not understand from the assigned reading. As the students come to class with the text material having been read, I go over any example from the text they did not understand and then add a few more as needed. Most of the information is presented in the text, so to lecture what they have already read is somewhat boring and redundant. I try to make use of some of the in-class project ideas to supplement the examples being shown. Depending on the class and the pacing, I may have students turn in a few problems before they leave each class day.

ACKNOWLEDGMENTS

I would like to thank all the students who have made this book possible. My students at Broome Community College in Binghamton, NY, suffered through the first conception of Jack problems and provided some great feedback. I promise that any future paintball examples will be much more realistic! My students at Doña Ana Community College in Las Cruces, NM, were the first to see Jack in this format and they also provided some great feedback. Overall, the students, staff, and faculties of New Mexico State University, Broome Community College, Doña Ana Community College, and the University of Texas at El Paso have been very encouraging of my creative approaches to mathematics in the classroom. Every student who has finished a semester with me has had a hand in the development of this project, and even a few who didn't make it that far.

The creative team at Cognella has been invaluable in this process. Marissa Applegate made first contact and her help in getting this project realized must be recognized. Thank you Marissa! The editorial staff has been great in working with me as a first-time author and helping to make sure everything was both as I wanted and in a format that would work. The graphic design and promotions departments did an amazing job as well.

My family has always consisted of my biggest fans, from parents and siblings to aunts, uncles, cousins, grandparents, nieces, and nephews. It is with their help and encouragement that I can extend myself each day to try something new and be better at my job. Special thanks to Sunny, Kasey, and Alex for allowing me the time to do this even when it took away a bit of our family time. Your help with character names and other plot ideas kept the project rolling. You have my eternal love and gratitude.

Tuesday J. Johnson

CHAPTER ONE

FINANCES

1.

Our story begins with Jack as a 13-year-old really wanting an Xbox Kinect to play some of his favorite games. He has played these games before with his friends, but he wants to be able to play at home so he can get better. Jack makes only $10 a week in allowance so it will take him a while to save up the total of $499.96 plus tax for the sweet Xbox Kinect package he found at Best Buy last time he was there. He has checked his savings account and he has only $214.56 saved up from birthdays, Christmas, and allowance. He musters up his courage and sits down with his parents to talk about a loan. His dad agrees that a loan is an acceptable way for Jack to reach his goal. His dad offers to lend Jack the remainder of what he needs at a simple interest rate of 2% for one year. That is, he will pay interest and it must all be paid off one year from the date of the loan.

1.1. If the tax rate in his community is 8.125%, find the price Jack must pay for his new system.

1.2. Knowing what he needs to pay and what he has available, find the amount that Jack is borrowing from his parents.

1.3. Considering the simple interest loan terms and the amount of the loan, determine how much Jack must pay back one year from the date of the loan.

1.4. Is this a reasonable amount for a 13-year-old to repay in one year's time?

Jack loves his Xbox Kinect and feels proud that he was able to earn the money to pay for it himself. He learned a valuable lesson about loans and simple interest. A few years later, Jack was told by his grandmother that she put $5,000 in an account for him on the day he was born. She didn't mention it to him earlier because she was hoping that he would use the money for college. Being a curious guy, and suddenly seing $$$ in his mind's eye, Jack asked all the details about the account. His grandmother told him that he could have the money when he was 18 if he enrolled in college, but had to wait until age 25 to get the money if he didn't go to college. The account was opened when savings accounts paid an acceptable interest rate of 3.5% compounded monthly.

1.5. If Jack decides to go to college, how much money will the account have? How much of this is interest?

1.6. If Jack decides not to go to college and has to wait until he is 25, how much money will the account have? How much of this is interest?

1.7. Would it be worth leaving the money in the account and going to college at age 25 or should he just go immediately after high school at age 18?

2.

Deciding that college was the obvious choice, during his senior year of high school, Jack started looking into how he was going to pay for college. He read as much as he could and learned that he would have to fill in a FAFSA (Free Application to Federal Student Aid) form and have the results sent to his school of choice. While filling in the form, he needed to ask his parents for their tax information. He had heard of them complain about taxes before but didn't really know what they were all about. He decided to put that on his list of things to learn about another time. After submitting the form electronically and applying to a few schools, Jack had to play the waiting game.

One sunny spring day Jack's wait ended. He came in the door after school and saw mail addressed to him from Tensed Technical University, TTU was his first choice of schools to attend and he saw that the envelope was fat!! Everyone knew that a skinny envelope was a rejection letter but fat envelopes had so much more information to help you get started at that school. He screamed and yelled and did a little dance, which startled his little sister. Jack had to help calm Mary before running to find his mom to tell her the great news. Later that night, after dinner, Jack and his parents read over the brochures and learned that it would cost $9,534 per semester for tuition since he was an out-of-state student. He knew that he would also have to pay for a room in a residence hall and a meal plan, which cost $1,973 and $1,618, respectively, per semester. They also estimated the cost of books to be $600 each semester. This college thing sure was sounding expensive! With scholarships, grants, and loans, the three of them agreed that it was ok to accept the offer to attend TTU.

1.8. Calculate the cost for Jack to attend TTU for one full school year. Add in about $100 per month in spending money.

1.9. Assuming Jack has scholarships and grants that will cover 43% of his costs, determine how large of a loan Jack needs to cover the rest of his expenses. If Jack is able to earn his desired degree in 5 years, what is the total of Jack's student loan burden?

1.10. Jack qualifies for a subsidized federal student loan with an interest rate of 6% and a repayment period of 20 years, beginning 6 months after he leaves college. A subsidized loan means that the government will pay the interest while he remains in school, but he has to start paying the interest once the repayment period begins. Using the total amount of the loan Jack needs from problem 1.9, determine how much each monthly payment will be.

Freshman orientation was a blast! This school really had a party atmosphere, which Jack loved. He was used to a small town so he was excited to be in the big city, all on his own, without having to listen to his parents' constant lectures about being responsible. He was a college man now! He could make all his own decisions. At one of the booths during orientation he saw that if he signed up for a credit card he could get a $50 gift card to the bookstore. Knowing a deal when he saw it, Jack signed up immediately. Since he was a college student Jack was already pre-approved and received the card in the mail two weeks later with a $1,000 credit limit and APR of 23%. Score!! He decided to take his friends out to eat and then spent the weekend shopping for a Nintendo 3DS XL with some games, a new jersey of his favorite NFL team, and some cool new shoes. The next month, Jack's bill came in the mail. Uh oh. He didn't have the $637.29 to pay the card off completely, but he did notice that all he had to pay was a minimum payment of $30.00 each month. That was something he could afford, so he wrote the check and sent it out the next day.

1.11 While out with his friends at dinner, Jack needs to show that he's a good guy and has to leave a tip. The bill came to $77.43 (they were all too young to drink alcohol) and he wanted to leave an 18% tip. How much should Jack add to the bill? What amount will show up as the total on Jack's credit card bill?

1.12. If Jack does not make any more purchases, use a table to determine how long it will take him to pay off the total, making only the minimum payment each time. How much does he repay in total? Hint: The table has been started for you here (you may not need all the lines):

Month	Amount Owed	Interest	Payment	New Amount
1	607.29	607.29(.23/12)=11.64	−30.00	588.93
2	588.93	588.93(.23/12)=11.29	−30.00	570.22
3				
4				
5				
6				
7				
8				
9				
10				
11				
12				
13				
14				
15				
16				
17				
18				
19				
20				
21				
22				
23				
24				
25				
26				
27				
28				
29				
30				

3.

After having an amazing college experience (see other chapters for some of his activities), Jack earns his degree in Architecture and is headed off into the "real world" to meet new challenges.

Jack has so many interviews before he is offered a job that he starts to get worried that all that time in college wasn't worth it! He had met a beautiful girl, Diane, while at TTU and they were engaged to be married soon. He needed a job so he could support his fiancée and they could start their lives together. Though he was offered several jobs, the pay wasn't enough to cover their budget each month.

1.13. Jack and Diane are both 23 years old. They each bought their cars used, and have gym memberships, student loans, and some credit card debt. They want to rent a two-bedroom apartment to start their life together. Determine an acceptable monthly budget for these two.

1.14. With the budget from problem 1.13, determine how much Jack needs to be paid each month to cover half the budget and to have some money to set aside for a future down payment on a house.

Just when Jack is about to give up his dream and go to work for his dad, a company calls and offers Jack a job with excellent benefits. Feeling as though he has dodged a bullet, Jack jumps at the opportunity and signs a contract. Excited to be making $60,000 a year, Jack looks forward to receiving his first paycheck. He knows that $60,000 paid out over 12 months should be $5,000 a month, but he finds a lot less in the check total. He's confused as he starts looking over the pay stub, finding abbreviations taking all of his money. That's when Jack remembers his dad always complaining about the government taking their part before he got his. So this is what taxes are all about. Bummer.

1.15. Assuming gross pay of $5,000 in his monthly check, determine the amount of his take-home pay. (Use 11.8% for federal withholding, 5.8% for social security, 6.4% for retirement, and 13.9% for health insurance.)

1.16. One of Jack's benefits is that his employer matches his retirement contributions for each paycheck. Determine the total amount paid in to Jack's retirement each month.

1.17. If Jack keeps this job until he retires at the age of 65, with the same amount being put into his retirement fund each month, and with an interest rate of 4.2%, how much money will be in the account when Jack retires?

4.

Jack and Diane have been married for three years now and are both settled in their careers. They decide that it is time to start a family and the best way to do that is to buy a house. Before they find their perfect home, Diane tells Jack that she is pregnant … with twins!! Jack is excited by the news but starts sweating a bit. There is no way they can stay in this two-bedroom apartment; they need to hurry their search for their new home. Together they decide they need to have a four-bedroom home with three bathrooms. After looking every day for two weeks, they finally find the perfect house.

1.18. The house they found costs $205,400. They have managed to save $40,000 for a down payment but will need a loan for the remainder of the balance. How much will they need to finance?

1.19. If Jack and Diane have decent credit and can get an interest rate of 3.3% on a 30-year home loan, what will be the amount of their monthly payment?

1.20. If they can get a better rate of 2.7% for a 15-year loan, what would be the amount of the payment?

1.21. Compare and contrast the two loans in 1.19 and 1.20; which is the better deal on a monthly basis? How much will they end up paying in total for each of the two loans? Which loan is the better deal in the long term?

They still have the same cars they had back when they were college students, but with twins on the way, it will be hard to get the kids into and out of their beat-up old cars. They flip a coin and decide that Diane gets to have a new car that will hold all four of them, while Jack keeps his old one for a few more years.

1.22. Look online or in dealership magazines to find a car, truck, or SUV that can accommodate this growing family. What is your vehicle choice and what does it cost?

1.23. If Jack and Diane still have good credit and can get a car loan for 5.2% monthly for 6 years, how much will the payments be on their new automobile? How much will they pay in total for this vehicle?

Life is going along pretty well for Jack, Diane, and the twins, Nicole and Sally. The parents have been making steady progress in their careers with promotions and recognition for excellent work, while the twins have been growing up playing sports, participating in band, and keeping good grades. The twins are now seniors in high school and are applying to colleges. Though Jack and Diane have great memories of TTU, they are doing their best not to pressure the twins into going to their alma mater. As they start the application process, Jack runs into the dreaded FAFSA again. This time, it is his taxes that must be entered.

1.24. For the 2012 tax year, Jack and Diane file their taxes as married filing jointly. They have the twins as dependents, give $2,000 a year to their favorite charity, and pay $13,440 into an IRA. They own their home and paid $5,482 of interest as well as student loan interest of $1,767. Determine if Jack and Diane will take the standard deduction or if they should itemize. If they make $145,000 combined each year, determine the amount of their taxes. (Refer to figures 1B and 1D as necessary.)

With the twins in college and the house to themselves, Jack and Diane are enjoying their (mostly) empty nest. They both have good cars, the house is paid off, and their careers are rolling along well. Just when Jack didn't think things could get any better, a financial recession hits the country and his job is downsized. Having a high salary, the company has to lay him off with a small severance package of $15,000, but he gets to keep his retirement. Diane has a recession-proof job as a paralegal so she will continue working. They decide that they can wait a few more years before drawing on Jack's retirement so he takes up a few hobbies and becomes a househusband.

The twins have finished college and married. Diane has reached a point she is ready to retire along with Jack so they can spend more time with their grandkids.

1.25. If Jack has $425,398.22 in his retirement account with an interest rate of 4.2% compounded monthly, how much will he receive per month if he starts withdrawing over a 25-year period of time?

1.26. If Diane has $385,729.13 in her retirement account with an interest rate of 3.9% compounded monthly, how much will she receive per month if she starts withdrawing over a 20-year period of time?

With their house paid off, cars all running well, and a comfortable retirement fund paying the remaining bills each month, Jack and Diane spend their days sitting together talking about their incredible life.

1.

SALES TAX

In the state of New Mexico in 2013, the state sales tax was 5.125%. However, you may pay up to 8.625% at your store due to local municipalities' taxes added in. Other states, such as Delaware and New Hampshire, have no state sales tax and do not allow municipalities to charge any either. Alaska, Montana, and Oregon have no state sales tax, but local counties and cities are allowed to set their own tax rates.

To compute sales tax we use the following formula:

(total cost of item) × (percent of sales tax as a decimal) = amount of sales tax

Once we have the amount of sales tax, we can find the register price of an item by adding the total cost of the item and the amount of sales tax.

It is important to note the use of sales tax in decimal form rather than in the percentage given. A percentage is an amount out of 100 (per = for every, cent = 100). This means that 8% is 8 for every 100. If we write this as a fraction, we have $8\% = \dfrac{8}{100} = 0.08$. This is the strictly mathematical view and is the most complete way of viewing percentages. However, think about the U.S. monetary system. We would write 8 cents as $0.08. In a practical application 8% is 8 pennies for every 100 pennies, also known as 8 cents on the dollar. When I rewrite percentages in decimal form, I always think of how to write it as money.

Example 1: You find the perfect throw rug at a local discount store for $19.88. You are in an area that charges 6% sales tax. How much will you pay for the rug?

Example 2: You see an HP laptop at Staples that you just have to have. The list price is $899.99. If the sales tax rate is 7.25%, how much will you pay for your new computer? (Ignore extra warranty protection plan options.)

Solution 1: The amount of sales tax is 19.88(0.06) = 1.1928. However, this will be rounded to the nearest cent, for total sales tax of $1.19. The total amount paid for the throw rug is 19.88 + 1.19 = $21.07.

Solution 2: The amount of sales tax is 899.99(0.0725) = 65.25 (rounded to the nearest cent). The total amount paid for the computer would be 899.99 + 65.25 = $965.24.

SIMPLE INTEREST

Simple interest is just an extension of sales tax. The simple interest formula can be represented by
Simple Interest = (Principal) × (Rate, as a decimal) × (Time, in years)

If we consider a simple interest loan, the principal is the amount of the loan. Just as with sales tax, the rate is always converted from a percentage into decimal form. For this formula, we use time in years; this means that if the loan is for a certain number of months, we must convert into years.

Example 3: I borrow $500 for 1 year at a simple interest rate of 4%. How much interest accumulates over that time? How much would I have to repay in total?

Example 4: I lend $4,000 to my daughter with the condition in must be repaid in 18 months. If I charge 2.5% interest, how much would she owe at the end of the time period?

Solution 3: The principal is $500, the rate is 4% = 0.04, and the time is 1 year. Our formula would be simple interest = 500(0.04)(1) = 20. This means I would owe $20 in interest and would have to repay a total of $520.

Solution 4: The principal is $4,000, the rate is 0.025 and the time is $\frac{18}{12} = 1.5$ years. (We divide by the 12 months that are in one year.) We now have 4000(0.025)(1.5) = 150. That is, the interest comes to $150 with a total repaid of $4,150.

Simple interest deals with more than just loans; you can earn money through simple interest deposits as well.

Example 5: Suppose you invest $1,000 in a savings bond that pays 10% simple interest per year. How much total interest will you receive in 5 years?

Example 6: You deposit $2,500 in an account that pays you simple interest of 3.2% per year. How much will you have in the account after 39 months?

Solution 5: We have $1000(0.10)(5) = \$500$ in interest after the five years.

Solutions 6: We will have a total of $2500 + 2500(0.032)\left(\dfrac{39}{12}\right) = 2500 + 260 = \$2760$ in the account after 39 months.

The solution method used in example 6 gives us a glimpse of another formula used for simple interest. This is the formula to compute future value with simple interest.

$$\text{future value} = \text{principal} + \text{interest}$$

or

$$\text{future value} = \text{principal} + \text{principal(rate)(time)}$$

COMPOUND INTEREST

Compound interest is interest that accrues on the initial principal and the accumulated interest of a principal deposit, loan, or debt. In example 5, we earned interest on only the original $1,000 each year. This is thought of as a pay-out to the investor as a dividend at the end of the year and only the $1,000 remains in the account as the next year starts. Compound interest is what occurs if the interest for the first time period is put back into the account so that there is now a larger principal for the second time period, and so on. This is how you earn interest on your interest, hence it is compounding. Compounding of interest allows a principal to grow at a faster rate than with simple interest.

According to Investopedia.com, "The more frequently interest is added to the principal, the faster the principal grows and the higher the compound interest will be. The frequency at which the interest is compounded is established at the initial stages of securing the loan (or opening the account)."

 What is compound interest?

The compound interest formula:

$$A = P\left(1 + \frac{r}{n}\right)^{nt}$$

where A is the total amount accumulated after t years, P is the starting principal, r is the annual percentage rate as a decimal (also known as APR), t is the number of years, and n is the number of compounding periods per year. The compounding periods will be given by words such as annually, quarterly, monthly, weekly, daily, and so on.

Example 7: Suppose I invested $10,000 at an APR of 4% for 10 years with annual compounding. How much would I have at the end of the term? How much of this is interest?

Example 8: Suppose I invested $10,000 at an APR of 4% for 10 years with weekly compounding. How much more would I have than with annual compounding?

Example 9: You invest $2,000 for 5 years with an APR of 3% and daily compounding. How much interest have you earned at the end of the term?

Solution 7: Using the formula we have $A = 10,000\left(1 + \dfrac{.04}{1}\right)^{(1\cdot10)} = 10,000(1.04)^{10} = \$14,802.44$.
Since we started with $10,000, we know that $4,802.44 is the amount of interest.

Solution 8: Again the formula will tell us the future amount:
$A = 10,000\left(1 + \dfrac{.04}{52}\right)^{(52\cdot10)} = \$14,915.95$. This amount is $14,915.95 - 14,802.44 = \$113.51$ more than with annual compounding.

Solution 9: In order to find interest, we first compute the future amount.
$A = 2000\left(1 + \dfrac{.03}{365}\right)^{(365\cdot5)} = \$2323.65$. Since $2,000 was invested at the start, this means we have accumulated $323.65 in interest.

ANNUAL PERCENTAGE YIELD

We can see that different compounding periods lead to different future amounts, even when the APR is the same. In order to compare accounts, we use annual percentage yield (APY). The APY is the actual percentage by which a balance increases in one year. In the case of annual compounding, the APY is equal to the APR. If the interest is compounded more than once a year, the APY will be greater than the APR. It is important to note that the annual percentage yield does not depend on the starting principal and we sometimes refer to it as the effective yield or simply the yield. The annual percentage yield is given by the formula

$$APY = \frac{absolute\ increase}{starting\ principal}$$

where the absolute increase is the amount of interest earned from the account in the given period of time.

Example 10: A bank offers an APR of 5.5% compounded daily. Find the APY.

Example 11: A bank offers an APR of 6.25% quarterly. Find the APY.

Example 12: A bank offers an APR of 4.2% compounded monthly. Find the APY.

Solution 10: As the principal does not make a difference (try various starting principal amounts to verify this for yourself) we will use the easiest principal available, $1. The future amount on $1 at an APR of 5.5% compounded daily for one year would be $A = 1\left(1 + \dfrac{.055}{365}\right)^{365} = 1.056536$. To find the absolute interest, we subtract the original $1 investment then substitute into the formula in order to get $APY = \dfrac{0.056536}{1} = 5.65\%$

Solution 11: The future amount: $A = 1\left(1 + \dfrac{.0625}{4}\right)^{4} = 1.06398$. The absolute interest is 0.06398 and so our APY = 6.40%.

Solution 12: The future amount is 1.042818 with absolute interest 0.042818. This makes the annual percentage yield 4.28%.

2.

COLLEGE COSTS

Frequently, college costs are given on a per-semester basis. It is important to double this value to account for a full year of college costs. If your college uses the quarter system, you would need to multiply by the appropriate number of quarters for the school year. On top of the tuition and fees, books, and other ancillary costs associated with school, you also need to factor in a variety of monthly costs. These monthly costs could just be spending money (if you live in a residence hall and do not have a car or cell phone), or could be as extensive as rent, food, insurance, gas and repairs, cell phone, and spending money. All of these costs must be kept in mind when considering attending college. For many prospective students, college would not be possible without the help of a student loan. About two-thirds of all college students take out student loans, with an average debt of about $20,000 at graduation.

LOAN BASICS: SUBSIDIZED AND UNSUBSIDIZED

For any loan, the principal is the amount of money owed at any particular time. Interest is charged on the loan principal. To pay off a loan, you must gradually pay down the principal. Therefore, in general, every payment must include all the interest you owe plus some amount that goes toward paying off the principal. There are two main types of student loans: Subsidized and Unsubsidized Student Loans.

A subsidized student loan is a loan that does not require you to pay interest while you are enrolled in school. During that time, the federal government pays the interest. However, after you graduate and your grace period (usually 6 months) ends, you must start paying back your loans and interest. Subsidized loans are based on financial need. The subsidized Stafford Loan and the Perkins Loan are classified as subsidized loans.

An unsubsidized student loan is a loan that requires you to pay back the interest on the loan while you are in school. Like a subsidized student loan, payment on your principal is deferred until six months after graduation, but instead of the school or government picking up the tab on the interest, it is all up to you.

On the surface it seems the only difference between subsidized and unsubsidized student loans is who pays the interest while you are still in school. There are some other major differences, as well. Subsidized loans have a tight cap on how much you can borrow per year and are dependent upon your specific situation and financial status. An unsubsidized loan also has a cap on it, but it is much higher than the subsidized loan. Basically, you can borrow between $4,000 and $5,000 more per year during your undergraduate career. When you have reached the cap on borrowing money through a subsidized loan, the only other option is an unsubsidized loan. It is very possible that you will end up with a combination of the two.

INSTALLMENT LOANS

A loan that you pay off with equal, regular payments is called an installment loan (or an amortized loan). This is the type of loan that is offered most in our society and includes home loans, car loans, and student loans. The loan payment formula is

$$PMT = \frac{P \cdot \left(\frac{r}{n}\right)}{\left[1 - \left(1 + \frac{r}{n}\right)^{-nt}\right]} .$$

For this formula, P is still the principal (the current amount of the loan), r is still the APR in decimal form, n is the number of payment periods per year, t is time of the loan in years, and PMT is the amount of payment to repay the loan. The numerator of this formula will tell you the amount of interest you owe for any given time period. The formula combined will give the amount of interest along with the portion of the principal that you are paying in each payment.

Example 13: You have a total of $50,000 in student loans with a fixed APR of 6% for 20 years. How much are your monthly payments? How much did you repay in total for this loan?

Example 14: You borrow $12,500 over a period of 5 years at an APR of 12%. How much are your monthly payments? How much did you repay in total for this loan?

Solution 13: Substituting in the formula: $PMT = \dfrac{50,000\left(\frac{0.06}{12}\right)}{\left[1 - \left(1 + \frac{0.06}{12}\right)^{-12 \cdot 20}\right]} = \dfrac{250}{\left[1 - (1.005)^{-240}\right]} = 358.22$.

That is, our payments will be $358.22 each month for twenty years. To find the total amount that is repaid on this loan, multiply the amount of each payment by the total number of payments (notice that this value is your exponent: monthly for 20 years). The total amount repaid is 358.22(240) = $85,972.80.

Solution 14: Once again the formula gives: $PMT = \dfrac{12{,}500\left(\frac{0.12}{12}\right)}{\left[1-\left(1+\frac{0.12}{12}\right)^{-12\cdot5}\right]} = 278.06$. We will make

monthly payments of $278.06, which makes the total repaid 278.06(60) = $16,683.60.

Example 15: You buy a house with a home loan of $150,000 over a period of 30 years at a fixed APR of 4%. How much are your monthly payments? How much will you repay in total for this loan?

Example 16: Your student loan total is $24,000 with a fixed APR of 8% for 15 years. How much are your monthly payments? How much will you repay in total for this loan?

Solution 15: Our payment is $PMT = \dfrac{150{,}000\left(\frac{0.04}{12}\right)}{\left[1-\left(1+\frac{0.04}{12}\right)^{-12\cdot30}\right]} = \$716.12$ with total repaid of $257,803.20.

Solution 16: Our payment is $PMT = \dfrac{24{,}000\left(\frac{0.08}{12}\right)}{\left[1-\left(1+\frac{0.08}{12}\right)^{-12\cdot15}\right]} = \$229.36$ with a total repaid of $41,284.80.

The portions of installment loan payments going toward principal and toward interest vary as the loan is paid down. Early in the loan term, the portion going toward interest is relatively high and the portion going toward principal is relatively low. As the term of the loan proceeds, the portion going toward interest gradually decreases and the portion going toward principal gradually increases.

CREDIT CARDS

Credit cards are different from installment loans in that you are not required to pay off your balance in any set period of time. Instead, you are generally required to make only a minimum monthly payment that typically covers all of the interest but very little principal. As a result, it takes a very long time to pay off your credit cards if you make only the minimum payment. If you wish to pay off your credit cards in a certain amount of time, you should stop using them and then use the loan payment formula to calculate the necessary payments.

For more of the history of credit cards and other useful information, please view this 56-minute video.

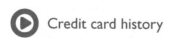 Credit card history

Example 17: You have a credit card balance of $2,300 with an annual interest rate of 21%. You decide to pay off your balance over 1 year. How much will you need to pay each month? (Assume you make no further credit card purchases.)

Example 18: Your New Year's resolution is to stop using your credit card and get it paid off. You have $5,000 on the card with an APR of 18% and you want to pay off the balance in one year. How much will you need to pay each month? How much will you have repaid in total on this card?

Solution 17: Using the loan payment formula, $PMT = \dfrac{2300\left(\frac{0.21}{12}\right)}{\left[1-\left(1+\frac{0.21}{12}\right)^{-12 \cdot 1}\right]} = \$214.16$.

Solution 18: Calculate the payment, $PMT = \dfrac{5000\left(\frac{0.18}{12}\right)}{\left[1-\left(1+\frac{0.18}{12}\right)^{-12 \cdot 1}\right]} = \$458.40$. To find the total repaid, we multiply the payment by twelve, the total number of payments made: $458.40(12) = \$5500.80$.

Example 19: Your card has an APR of 20% and you want to pay the balance of $21,656 off in three years. If you stop using the card immediately, how much will you need to pay each month? How much will you have repaid in total on the card?

Example 20: Your card has an APR of 12.5% and you want to pay the balance of $3,000 off in three years. Assume no further charges are made. How much will you need to pay each month? If you paid it off in one year instead, how much would you need to pay each month? Compare the totals repaid and determine which is a better idea.

Solution 19: First we find the payment: $PMT = \dfrac{21,656\left(\frac{0.20}{12}\right)}{\left[1-\left(1+\frac{0.20}{12}\right)^{-12 \cdot 3}\right]} = \$804.81$. To find the total repaid, we multiply the payment by the 36 times we made the payment. The total repaid is $28,973.16.

Solution 20: To pay off the balance in three years, your payments should be $PMT = \dfrac{3000\left(\frac{0.125}{12}\right)}{\left[1-\left(1+\frac{0.125}{12}\right)^{-12 \cdot 3}\right]} = \$100.36$. If we paid it off in one year instead, our payments would be

$PMT = \dfrac{3000\left(\frac{0.125}{12}\right)}{\left[1-\left(1+\frac{0.125}{12}\right)^{-12 \cdot 1}\right]} = \$267.25$. The payments to pay off the debt in one year are much higher and this can be too much for some people to handle. However, over the course of three years you would pay 100.36(36) = \$3,612.96 in total for your \$3,000 credit card total. Whereas if you pay \$267.25, for one year you would only pay 267.25(12) = \$3,207 in total. This is a savings of over \$400.

Credit cards can be a good thing when used correctly. They can also turn into a nightmare when used poorly. In order to avoid credit card trouble, you should use only one credit card. When you have balances on many cards, you lose track of overall debt. Also, if you lose your purse or wallet, you will have fewer cards to cancel and replace. If possible, pay off your balance in full each month. Using the card in this way may earn you valuable rewards. If you plan to pay off your balance each month, make sure your card has an interest-free grace period. This allows you to use the purchasing power of a credit card interest free, paying off all the purchases at one time.

Fees and rates differ greatly among credit cards. At this point in time, there is no reason to have a credit card that charges an annual fee. The interest rate is the most important thing to consider when looking at credit cards. When first signing up for a credit card, be careful of teaser rates. These are low rates that are offered for a short period of time, usually 6 months, after which the card reverts to very high rates. You should use your credit card for cash advance only in case of an emergency. Nearly all cards charge both fees for access and high interest rates. In addition, most credit cards charge interest immediately on cash advances, even if there is a grace period on purchases. When you need cash, get it directly from your own bank by cashing a check or using an ATM card. The checks credit card companies send in the mail are cash advance checks; these are a bad, bad idea. If you own a home, consider replacing a common credit card with a home equity credit line. You will generally get a lower interest rate, and the interest may be tax deductible. If you ever get in trouble with credit card debt, never fear consulting a financial advisor.

3.

BUDGETS AND CONTROLLING YOUR FINANCES

In order to have a good control of your finances, you should keep four things in mind:

- Know your bank balance.
- Know what you spend.
- Do not buy on impulse.
- Make a budget.

We have all heard the advice to make a budget and keep to it, but what is a budget? Setting a budget just means that you know how much you spend, how much you earn, and that you make adjustments so that you do not end up in debt. To set up your budget, first list all your monthly income. Be sure to include any prorated amount—that is, what it averages out to per month—for any income you do not receive monthly (such as student loan income). Next, list all your monthly expenses. In this list include prorated amounts as well. Do not forget tuition, books, car insurance, vacations, holiday gifts, and so on. The third step is to subtract your total expenses from your total income to determine your net monthly cash flow. If this number is positive, you are on the right track. If this difference is negative, you should make adjustments as necessary. A negative monthly cash flow will place you firmly in debt.

Some expenses to consider when creating your budget include, but are not limited to:

- Mortgage/rent
- Transportation—gas, insurance, oil, tires, registration, bus tokens, subway tickets
- Utilities—gas, electric, trash, water, sewer
- Food and Necessities—personal hygiene, grooming, clothes, cleaning supplies, groceries

- Extras—cable/satellite television, cell phone, home phone, internet, gym membership, child care, pet care
- Government—taxes, health insurance, lawyer fees, fines

Example 21: Compute the total cost per year.
- a) Maria spends $20 every week on coffee and spends $130 per month on food.
- b) Suzanne's cell phone bill is $85 per month, and she spends $200 per year on student health insurance.
- c) Vern drinks three 6-packs of beer each week at a cost of $7 each and spends $700 per year on his textbooks.

Solution 21:
- a) Coffee is $20 each week for 52 weeks, which is $1,040 per year. Food is $130 per month for 12 months, which is $1,560 per year. Maria spends a total of $2,600 per year.
- b) The cell phone costs $85 per month for 12 months, which is $1,020 per year. Her health insurance is already on an annual basis, so Suzanne spends $1,220 per year on these items.
- c) Three 6-packs per week is $21 per week. Over the course of 52 weeks, Vern's beer will cost him $1,092. Add to this his $700 on books and Vern spends $1,792 per year.

Example 22: Prorate the following expenses and find the corresponding monthly expense.
- a) Luisa pays $5,600 for tuition and fees, plus $400 for textbooks, for each of two semesters at college.
- b) Ian pays a semiannual premium of $650 for automobile insurance, a monthly premium of $125 for health insurance, and an annual premium of $400 for life insurance.
- c) Randy spends an average of $25 per week on gasoline and $45 every three months on the daily newspaper.

Solution 22:
- a) Luisa pays $6,000 each semester for two semesters. This is a total of $12,000 per year. To prorate this amount into a monthly expense, we divide by 12 months in a year to get $1,000 per month.
- b) Ian will pay $1,300 for automobile insurance, $1,500 for health insurance, and $400 for life insurance per year. Adding these values and dividing by 12 will give a monthly expense of $266.67.
- c) Each year, Randy spends $1,300 on gasoline and $180 on the newspaper. This total of $1,480 prorated to a monthly cost is $123.33.

Example 23: Which option is less expensive? You currently drive 400 miles per week in a car that gets 26 mpg. You are considering buying a new fuel-efficient car for $18,000 that gets 48 mpg. Insurance premiums for the new and old car are $800 and $400, respectively. You anticipate spending $1,500 on repairs for the old car each year and having no repairs for the new car. Assume gas costs $3.24 per gallon. Over a five-year period of time, is it less expensive to keep your old car or buy the new car?

Solution 23: First, let us calculate the cost of operating the old car for a five-year period of time. Driving 400 miles per week for 52 weeks per year for 5 years is 104,000 miles. The old car gets 26 mpg so it will need $104,000 \div 26 = 4000$ gallons of gas at $3.24 per gallon. You will spend a total of $12,960 on gas for the five years, plus $400(5) = $2,000 for insurance and $1,500(5) = $7,500 for repairs. This leads to an expense of $22,460 for five years for driving the old car each year. We use the same reasoning to calculate $104,000 \div 48 = 2166.67$ gallons of gas costing $7,020 for the five years. Add in $800(5) = $4,000 for insurance along with the price of the car and the new car will cost $29,020 over the five-year period of time. Though the improved gas mileage sounds like a great deal, and you may really want a new car, it is more economical to stick with the old car in this situation.

Example 24: You could take a 15-week, three-credit college course, which requires 10 hours per week of your time and costs $300 per credit-hour in tuition. Or during those hours you could have a job paying $10 per hour. What is the net cost of the class compared to working? Given that the average college graduate earns nearly $20,000 per year more than a high school graduate, is paying for the college course a worthwhile expense?

Solution 24: The class will cost $900 in total for three credits at $300 per credit. Working 10 hours per week for 15 weeks at $10 per hour will result in pay of $1,500. You would have to spend $900 and miss out on $1,500 in wages, leaving a net cost of $2,400 for this class. This seems to be a relatively small amount when you consider the $20,000 per year more you could earn for the remainder of your working life.

PAYCHECKS

Once you enter the workforce, you are probably most excited about getting a paycheck. You know how much you earn per hour and how many hours you worked, so why isn't that the amount in your check? You should expect your employer to withhold federal income tax, Social Security tax, and Medicare tax from your paychecks. The federal government regulates these deductions and requires employers to withhold them, unless the employee is exempt. Your employer must withhold federal taxes according to the federal government's instructions. Specifically, she withholds federal income tax based on your W-4 form and the Internal Revenue Service Circular E's tax-withholding tables. Circular E also has the tax rates for Social Security and Medicare taxes, which are based on flat percentages of your pay. Similar to federal taxes, individual states may or may not withhold income taxes from your paycheck. States such as Texas and Washington do not have state income tax; however, their neighbors New Mexico and Idaho do. This can cause some difficulties for individuals residing in a state with income tax and working in a state that does not.

Though most states do not require state disability insurance and unemployment tax withholding, if they apply, your employer must withhold them according to the administering agency's rules. For example, as of 2011, Pennsylvania, New Jersey, and Alaska require employers to withhold state unemployment tax from employees' paychecks, and states such as California, Rhode Island, and New Jersey allow disability insurance withholding. Employers generally offer employees some type of voluntary benefit, which may include medical and dental plans, short-term and long-term disability, flexible spending accounts, and 401k plans. Provided your employer has an established company policy and you accepted the benefit, these deduction are allowed. Voluntary deductions may also include paycheck advances or loans, which you may repay in installments. You can stop voluntary deductions at any point. For example, if you are no longer enrolled in your company's health or retirement plan, your employer should stop the deductions.

Terminated employees should check with their state department of labor for deductions that an employer can make from the final wages. The state might also have specific deductions that it considers illegal, such as for uniforms, gratuities, and business expenses. A wage garnishment is a legal deduction but applies only to the employee whose wages are being garnished by a government entity or court order, as for child support.

A sample paycheck with deductions:

123—John R. Doe	Pay Period 06/02/06 to 06/16/06			Required Deductions		
Earnings				Federal Income Tax	00.00	00.00
Hours	Rate	This Period	YTD	FICA—Medicare	06.08	12.16
50	9.00	450.00	900.00	WI State Income Tax	00.00	00.00
Gross Pay		**450.00**	**900.00**	FICA—Social Security	25.92	51.84
				Other Deductions		
				Health Insurance	00.00	00.00
				401k	00.00	00.00
				Parking	00.00	00.00
				NET PAY	**$418.00**	**$836.00**

Your Employer
1234 Some Street
Milwaukee, WI ZIPCODE

Check Number: xxxxxx
Pay Date: 06/19/06

PAY   * * *Four hund eighteen dollars and 00 cents* *$418.00

To the Order of
     John R. Doe
     555 Some Street
     Milwaukee, WI ZIP CODE

Figure 1A: A sample pay check with stub.

RETIREMENT

If your employer offers a retirement plan, it is always a good idea to invest unless you already have a better account somewhere else. Why do we need a retirement account? Through the wonders of modern medicine, we are living to a life expectancy of 79 years in the United States. No one plans to work until the day they die, and so retirement accounts offer you the income you need after you have finished working. A retirement account is a form of an annuity that can also be called a savings plan. The savings plan formula is

$$A = PMT \times \frac{\left[\left(1+\frac{r}{n}\right)^{nt} - 1\right]}{\left(\frac{r}{n}\right)}$$

where A is the accumulated savings plan balance, PMT is the regular payment (deposit) into the account, r is the annual percentage rate (APR) in decimal form, n is the number of payment periods per year, and t is the number of years. (Take a moment to notice the similarities and differences between the savings plan formula and the loan payment formula.)

Example 25: At age 25 you set up an IRA (individual retirement account) with an APR of 5%. At the end of each month, you deposit $80 in the account. How much will the IRA contain when you retire at age 65? Compare that to the total deposits made over the time period.

Example 26: You put $200 per month in an investment plan that pays an APR of 4.5%. How much money will you have after 18 years? Compare that to the total deposits made over the time period.

Solution 25: From the time you start until you retire is a period of 40 years. Substituting appropriately into the formula we have $A = 80 \dfrac{\left(1+\frac{0.05}{12}\right)^{12\cdot40} - 1}{\left(\frac{0.05}{12}\right)} = 80(1526.020156) = \$122,081.61.$

We made a total of 480 deposits worth $80 each for a total of $38,400 in deposits. This means we earned $83,681.61 in interest.

Solution 26: The formula gives $A = 200 \dfrac{\left(1+\frac{0.045}{12}\right)^{12\cdot18} - 1}{\left(\frac{0.045}{12}\right)} = 200(331.8680176) = \$66,373.60.$

Compare this to our deposits totaling 200(216) = $43,200. Notice that the amount of growth in example 25 is much greater than the amount of growth in this example. The longer the money stays in the account, the more interest you will earn.

Example 27: Your goal is to create a college fund for your child. Suppose you find a fund that offers an APR of 5%. How much should you deposit monthly to accumulate $85,000 in 15 years?

Example 28: At age 22 when you graduate, you start saving for retirement. If your investment plan pays an APR of 4.2% and you want to have $5 million when you retire in 45 years, how much should you invest?

Solution 27: In this example, we know the accumulated amount but wish to find the payment. In order to do that, we will still compute the larger fraction portion of the formula, but then instead of multiplying in the last step, we will divide to find our solution.

$$85,000 = PMT \frac{\left(1+\frac{0.05}{12}\right)^{12 \cdot 15} - 1}{\left(\frac{0.05}{12}\right)}$$

$$85,000 = PMT(267.2889438)$$

$$\$318.01 = PMT$$

Solution 28: Similar to example 27 we find

$$5,000,000 = PMT \frac{\left(1+\frac{0.042}{12}\right)^{12 \cdot 45} - 1}{\left(\frac{0.042}{12}\right)}$$

$$5,000,000 = PMT(1599.303448)$$

$$\$3126.36 = PMT$$

With this example we find that you have to have a great job with excellent take-home pay, or a much better investment, in order to have $5 million in an account after working 45 years.

4.

THE AMERICAN DREAM—MORTGAGES

A home mortgage is an installment loan designed specifically to finance a home. A down payment is the amount of money you must pay up front in order to be given a mortgage or other loan. Closing costs are fees you must pay in order to be given the loan. They may include a variety of direct costs, or fees charged as points, where each point is 1% of the loan amount. In most cases, lenders are required to give you a clear assessment of closing costs before you sign for the loan.

The primary factors in determining loan payments are interest rate and loan term. Other factors include understanding the loan, down payment, fees and closing costs, and fine print. To understand the loan you need to know the length of the term of the loan, when your payments are due, if early payments are accepted, and if the interest rate is fixed or variable. For the down payment, you need to know if there is one and if so, how do you afford it? Will you be allowed to borrow money for the down payment, or do you have to show proof that you already have the money? If a down payment is not necessary, will you get a better interest rate by offering a down payment? Be sure you identify ALL closing costs, including origination fees and discount points, since different lenders quote their fees differently. Also, be sure to consider how these fees and costs affect the overall cost of the loan over its term. The fine print is what most people fear. This may make the loan more expensive than it seems on the surface. Be especially wary of any prepayment penalties, since you may later decide to pay off the loan early or to refinance at a better rate.

Offers to refinance your loan almost always seem enticing. Refinancing at a lower interest rate can save you money, but it may not be a good idea in all circumstances. First, consider how long it will take before your monthly savings cover the fees and closing costs you must pay for the new loan. As a general rule, you should not refinance if it will take more than two to three years to recoup.

Also consider that refinancing "resets the clock" on a loan. For example, if you have been paying four years on a ten-year loan, you have only six years remaining. If you refinance, you may be back to a new ten-year contract. This may not be worth refinancing.

The simplest type of home loan is a fixed-rate mortgage, in which you are guaranteed that the interest rate will not change over the life of the loan. Most fixed-rate loans have a term of 15 or 30 years. (The word "mortgage" comes from Latin and old French and means "dead pledge." It therefore should come as no surprise that these loans have long terms.) We use the same loan payment formula from the student loan discussion to calculate payments on home mortgages.

Example 29: You need a $200,000 home loan. Compare monthly payments and total loan burden for these two options.

- Option 1 is a 30-year loan at an APR of 8%
- Option 2 is a 15-year loan at an APR of 7%

Solution 29: We find the monthly payments just as we did in the student loan section. Starting with option 1 we have $PMT = \dfrac{200,000\left(\frac{0.08}{12}\right)}{\left[1-\left(1+\frac{0.08}{12}\right)^{-12\cdot30}\right]} = \$1467.53.$ If we pay this amount every month for 30 years, we will have a total loan burden of $1,467.53(360) = $528,310.80. Notice that this is more than two and a half times the loan principal. For option 2 we have $PMT = \dfrac{200,000\left(\frac{0.07}{12}\right)}{\left[1-\left(1+\frac{0.07}{12}\right)^{-12\cdot15}\right]} = \$1797.66.$ Option 2 definitely has the higher payments, however, the total loan burden is $1,797.66(180) = $323,578.80. This is a savings of over $200,000 compared to option 1.

Example 30: You need a $60,000 loan. Compare monthly payments and total loan burden.

- Option 1 is a 30-year loan at an APR of 7.15%
- Option 2 is a 15-year loan at an APR of 6.8%

Solution 30: Option 1 has monthly payments of $405.24 and a total burden of $145,886.40. Option 2 has monthly payments of $532.61 and a total burden of $95,869.80. Notice in both of these examples that the longer term loan has lower monthly payments but also has the higher overall burden. Both of these should be taken into consideration when deciding on your loan term.

Example 31: You need a $120,000 mortgage. Calculate the monthly payment and total closing costs for each option. Which would you choose?

a) Option 1 is a 30-year fixed rate of 8% with closing costs of $1,200 and no points. Option 2 is a 30-year fixed rate of 7.5% with closing costs of $1,200 and 2 points.

b) Option 1 is a 30-year fixed rate of 4.5% with no closing costs and no points. Option 2 is a 30-year fixed rate of 3.8% with closing costs of $1,200 and 4 points.

Solution 31: For each option, our principal is $120,000.

a) Option 1 has payments of $880.52 per month and closing costs of $1,200. Option 2 has payments of $839.06 per month but closing costs are $1,200 plus 2 points, or 2 percent of $120,000. That is, the closing costs are $1,200 + $2,400 = $3,600. The payments differ by only a little more than $40. The decision for this loan would come down to how much you could afford to pay up front.

b) Option 1 has payments of $608.02 with no closing costs or points. Option 2 has payments of $559.15. Once again the payments are different by only about $50. The big decision comes in whether or not you can afford the closing costs of $1,200 plus 0.04(120,000) = $4,800 for a total up front of $6,000. The lower payment will save nearly $18,000 over the life of the loan (saving about $50 each month for 30 years) compared to the $6,000 you would need to pay up front. However, if you do not have that money available to pay closing costs and points, you will need to make the higher monthly payments.

PREPAYMENT STRATEGIES

As the terms of a home mortgage are so long, the initial payments are almost entirely interest and very little principal. Every extra dollar that you can pay over your payment will be applied to the principal. That results in one less dollar that accrues interest in the long life of the loan.

Example 32: Suppose you have a loan of $117,500 at an APR of 4%.

a) What would be your monthly payments for a 30-year loan?

b) What would be your monthly payments for a 15-year loan?

c) If your loan is set up for 30 years and you wanted to pay it off in 15, how much extra should you pay each month?

d) Compare the total amounts paid in each loan.

Solution 32:

a) Your monthly payments would be $560.96.

b) Your monthly payments would be $869.13.

c) In order to pay extra each month in order to pay off the 30-year loan in 15 years, you should pay 869.13 – 560.96 = $308.17 extra each month.

d) If you stay to your 30-year plan, paying the minimum amount each month, you will end up paying a total of $201,945.60 for your loan. However, by accelerating your payments, you will pay a total of $156,443.40. It is well worth your money to pay extra each month if you can afford it within your budget.

ADJUSTABLE RATE MORTGAGES

A fixed-rate mortgage guarantees that your monthly payments (for the loan itself) never change. (Your payments may change due to insurance and taxes that have nothing to do with the loan itself.) A fixed-rate loan is bad for lenders if the interest rates rise. Lenders can lessen the risk of losing out by charging a higher APR for a longer-term loan. That is why 30-year loans generally have a higher APR than 15-year loans. An even better idea (for the lenders) is to offer an adjustable rate mortgage. With an adjustable rate mortgage (ARM), your interest rate will change based on the economy. You could have a nice home loan, paying $550 per month for several years, but then the market tanks and your interest rate changes and the bank now expects you to pay $1,800 per month for the same loan. Most ARMs include a rate cap, that is, a rate that the APR can never go higher than.

Example 33: You have a choice between a 30-year fixed-rate loan at 8.5% and a 30-year ARM with a first-year rate of 5.5%. Neglecting compounding and changes in principal, estimate your monthly savings with the ARM during the first year on a $125,000 loan.

Example 34: For the loan in example 33, suppose that the ARM rate rises to 10% at the start of the second year. Approximately how much extra will you then be paying over what you would have paid if you had taken the fixed-rate loan?

Solution 33: The 30-year fixed-rate loan has monthly payments of $961.14. For the first year, the ARM has monthly payments in the first year of $709.74. These payments are very tempting.

Solution 34: Over the course of the first year, not much is paid to the principal. For ease of computation, and not exactness, we shall guess that the amount of the principal is around $123,000 for the remaining 29 years of the ARM at 10% interest. This now brings the monthly payment to around $1,085.45. The $260 per month in savings by choosing the ARM to start has now turned into paying $120 more than the fixed-rate mortgage.

CARS

All that house buying seems to be complicated; is it just as bad for cars? Car buying is much like student loan shopping, only easier. Most car terms are now six years. This allows for less accumulation of interest, but there is still some in order to make the loan worthwhile to the lender. Knowing the loan-payment formula will help you walk into a dealership knowing what you can and what you cannot afford. If your budget allows only for car payments of $350 per month, you can determine the total cost of a car that you can afford. Similarly, if you see a car advertised, you can determine if you can afford the monthly payments. Being an informed consumer should always be your goal.

Example 35: A 2013 Dodge Challenger retails for $27,295. If you trade in your car and receive a $2,500 credit as down payment and are offered a 6-year loan with an APR of 5.7%, how much would you pay each month?

Example 36: Sunny wants to buy a Harley-Davidson Iron 883 that retails for $7,999. She wants to keep her current bike so she doesn't really have a down payment. How much would she have to pay each month for a 4.5-year loan at 5.1%?

Solution 35: First, we can subtract the down payment from the cost of the car to determine we need a loan of 27,295 − 2,500 = $24,795. The monthly payments for this loan would be $407.42.

Solution 36: With no down payment, Sunny will finance the full amount. Her monthly payments will be $166.09.

TAXES

For many people, taxes are straightforward enough that you could do them yourself. If you have an income with no investments or other means of making money, it could save you a lot of money in the long run. In order to start doing your own taxes, you need to know some tax basics.

Your gross income is all of your income for the year. This includes wages, tips, profits from a business, interest or dividends from investments, and any other income that you receive. Some of your gross income is not taxed, at least not right away. Tax-deferred savings plans such as IRAs fit into this category. You can contribute to an IRA tax-free but you will eventually pay taxes on it when you withdraw the money. These savings plans are considered adjustments to your gross income and so they are called adjusted gross income (AGI). That is, subtract your savings plan contributions from your gross income to get your adjusted gross income. Most people are entitled to certain exemptions and deductions as well. These are amounts that you subtract from the AGI before calculating your taxes. Once these are subtracted you have your taxable income.

A tax table, or tax rate computation, allows you to determine how much tax you owe on your taxable income. You may have tax credits (children) so you subtract the amount of any credits to find your total tax. If you look at your paystub, or W2 form, you will notice that you have already paid part of your tax bill through withholdings. Some people may be required to make estimated payments throughout the year as well. This generally applies to people who routinely owe a sum of more than $1,000 annually. Subtract the taxes that you've already paid to determine how much you may still owe, or in a good year, how much you will get back as a tax refund. There are several different computer programs that will assist you with doing your own taxes. You may also be able to file online for free through your state revenue website.

Example 37: Kasey earned wages of $34,200, received $823 in interest from a savings account, and contributed $1,400 to a tax-deferred retirement plan. She was entitled to a personal exemption of $3,800 and to deductions totaling $6,100. Find her gross income, her adjusted gross income, and her taxable income.

Solution 37: Kasey's gross income is the sum of all her income: Gross income = $34,200 + $823 = $35,023. Her contribution to a tax-deferred retirement plan is an adjustment, so her AGI is $35,023 − $1,400 = $33,623. To find Kasey's taxable income, we subtract her exemptions and deductions from her AGI. That is, AGI − Exemptions − Deductions = $33,623 − $3,800 − $6,100 = $23,723. Even though Kasey earned a gross income of $35,023, she needs to pay taxes on only $23,723 of it.

The actual amount of exemptions and deductions will vary depending on your filing status. There are four main statuses. Single is the status you select if you are unmarried, divorced, or legally separated. You would select married filing jointly if you are married (legally in your state) AND you and your spouse file a single tax return. That is, you include both your names and incomes on the same return. You select married filing separately if you are married (legally in your state) AND you and your spouse file two separate tax returns. The head of household status applies if you are unmarried and are paying more than half of supporting a dependent child or parent.

Personal exemptions are a fixed amount per person. In tax year 2012 this amounted to $3,800 for each individual in the household. The standard deduction varies based on your filing status. A table showing the most recent values is given in Figure 1B:

	2012	2011	2010	2009
Single	$5,950	$5,800	$5,700	$5,700
Married filing jointly	$11,900	$11,600	$11,400	$11,400
Married filing separately	$5,950	$5,800	$5,700	$5,700
Head of household	$8,700	$8,500	$8,400	$8,350
Personal Exemption	$3,800	$3,750	$3,650	$3,650

Figure 1B: Deduction and personal exemptions for various tax year.

Deductions vary from one taxpayer to the next. Many will use the standard deduction shown in the table. Others will choose to itemize; itemized deductions allow you to deduct interest from a mortgage, health expenses, going green, student loan interest, charity, and many other items. In order to itemize your deductions, you need to keep good records throughout the year.

Example 38: Should you itemize or take the standard deduction?
a) Your deductible expenditures are $8,600 for interest on a home mortgage, $3,000 for contributions to charity, and $645 for state income taxes. Your filing status entitles you to a standard deduction of $11,900.
b) Your deductible expenditures are $4,300 for contributions to charity and $760 for state income taxes. Your filing status entitles you to a standard deduction of $5,950.

Solution 38: a) Adding all possible deductions you have $8,600 + $3,000 + $645 = $12,245. This amount is larger than the standard deduction, so you would choose to itemize. b) Adding all possible deductions you have $4,300 + $760 = $5,060. As this is smaller than your standard deduction, you would choose to keep the standard deduction and not itemize.

The United States has a progressive income tax. This means that people with higher taxable incomes pay at a higher tax rate. The system assigns different marginal tax rates to different income ranges (or margins). Figures 1C–1F show the breakdown for various filing statuses in the tax year 2012.

If Taxable Income Is:	The Tax Is:
Not over $8,700	10% of the taxable income
Over $8,700 but not over $35,350	$870 plus 15% of the excess over $8,700
Over $35,350 but not over $85,650	$4,867.50 plus 25% of the excess over $35,350
Over $85,650 but not over $178,650	$17,442.50 plus 28% of the excess over $86,650
Over $178,650 but not over $388,350	$43,482.50 plus 33% of the excess over $178,650
Over $388,350	$112,683.50 plus 35% of the excess over $388,350

Figure 1C: Filing status of single.

If Taxable Income Is:	The Tax Is:
Not over $17,400	10% of the taxable income
Over $17,400 but not over $70,700	$1,740 plus 15% of the excess over $17,400
Over $70,700 but not over $142,700	$9,735 plus 25% of the excess over $70,700
Over $142,700 but not over $217,450	$27,735 plus 28% of the excess over $142,700
Over $217,450 but not over $388,350	$48,665 plus 33% of the excess over $217,450
Over $388,350	$105,062 plus 35% of the excess over $388,350

Figure 1D: Filing status of married filing jointly.

If Taxable Income Is:	The Tax Is:
Not over $8,700	10% of the taxable income
Over $8,700 but not over $35,350	$870 plus 15% of the excess over $8,700
Over $35,350 but not over $71,350	$4,867.50 plus 25% of the excess over $35,350
Over $71,350 but not over $108,725	$13,867.50 plus 28% of the excess over $71,350
Over $108,725 but not over $194,175	$24,332.50 plus 33% of the excess over $108,725
Over $194,175	$52,531 plus 35% of the excess over $194,175

Figure 1E: Filing status of married filing separately.

If Taxable Income Is:	The Tax Is:
Not over $12,400	10% of the taxable income
Over $12,400 but not over $47,350	$1,240 plus 15% of the excess over $12,400
Over $47,350 but not over $122,300	$6,482.50 plus 25% of the excess over $47,350
Over $122,300 but not over $198,050	$25,220 plus 28% of the excess over $122,300
Over $198,050 but not over $388,350	$46,430 plus 33% of the excess over $198,050
Over $388,350	$109,229 plus 35% of the excess over $388,350

Figure 1F: Filing status of head of household.

In addition to these marginal taxes, some income is subject to Social Security and Medicare taxes that are collected under the name Federal Insurance Contribution Act (FICA). Taxes collected under FICA help to pay those collecting Social Security and Medicare. FICA taxes apply only to income from wages and self-employment, not interest earned or from dividends. For tax year 2012, the FICA rate was 7.65%, which includes 6.2% to Social Security and the remaining 1.45% to Medicare health insurance.

Dividends and Capital Gains are incomes that get special treatment in the eyes of the tax collector. Capital gains have two main subcategories: short-term and long-term capital gains. A short-term capital gain applies to items sold within 12 months of their purchase, whereas a long-term capital gain applies to items held more than 12 months before being sold. In general, capital gains taxes are at a lower rate than standard income. A dividend is money earned on stocks and other investments.

Earlier, IRAs were mentioned as an adjustment to gross income. Though you can contribute to your retirement fund tax free, you will pay taxes when you begin receiving retirement account payments. At this time it goes from being an adjustment (pre-tax subtraction) to being an income (and therefore taxable). In collecting retirement we use the loan formula that we explored previously. This is because we have a set amount in the account that continues to accrue interest as we withdraw money until the account is down to zero (much like paying off a loan).

Example 39: Alexzander begins investing $75 per month in a retirement account paying 3.6% APR compounded monthly when he turns 22. He contributes to this account every month until he turns 55. How much does he have in the account when he stops contributing? Now Alexzander is 55 and he wants to make withdrawals that last the next 20 years. If the account rate of 3.6% APR compounded monthly holds during that time, how much will Alex earn from his retirement account each month?

Solution 39: We use the savings plan formula to determine how much will be in the account at the end of his contributions.

$$A = 75 \frac{\left(1 + \frac{0.036}{12}\right)^{12 \cdot 33} - 1}{\left(\frac{0.036}{12}\right)}$$

$$A = 75 \frac{(1.003)^{396} - 1}{0.003}$$

$$A = 75(758.2281966)$$

$$A = \$56,867.11$$

Now that Alexzander is 55, he begins withdrawing from this account and we use the loan payment formula to determine how much he should be paid each month in order to empty the account in 20 years.

$$PMT = \frac{56,867.11\left(\frac{0.036}{12}\right)}{\left[1-\left(1+\frac{0.036}{12}\right)^{-12\cdot20}\right]}$$

$$PMT = \frac{56.867.11(0.003)}{\left[1-(1.003)^{-240}\right]}$$

$$PMT = \frac{170.60133}{0.5127228178}$$

After contributing $75 per month for 33 years, Alex is able to withdraw $332.74 each month for the next 20 years.

IN- AND OUT-OF-CLASS PROJECT IDEAS

1. As a class, or in small groups, discuss items to include in a typical budget for a typical family of 3, 4, or 5. Include "everybody bills" and "some people bills."
2. Decide on a budget that you would expect to have five years after graduation, for the living situation you expect at that time.
3. Research what you can expect to earn in your chosen career about 5 years after graduation in the location you wish to live. Compare this with the individual budget to see if changes need to be made.
4. Search the internet or local sources to find a dream car to purchase. Determine the monthly payments for a given interest rate and period of time.
5. Search the internet or local sources to find a dream house to purchase. Determine the monthly payments for a given interest rate and period of time.
6. Go to a local bank or credit union to find the current rate for a savings plan. Determine the amount you would have in that account with monthly deposits of $100 for 18 years as a college savings plan for your child. Next, determine the amount in that account if you had monthly deposits of $250 for 40 years as a retirement fund. Be sure to include the name of the bank and the interest rate you were quoted.

CHAPTER TWO

NUTRITION AND EXERCISE

5.

Y ou may wonder how Jack managed to stay healthy all those years without ever getting sick and having to deal with doctors and hospitals and other medical expenses. Jack had one grandmother with diabetes and a grandfather with heart problems, so he was taught early about healthy living, including nutrition and exercise. Jack always tries to eat a well-balanced diet consisting of fresh fruits and vegetables, when possible, along with a variety of meats, breads, milk products, and only minimal amounts of sweets. He also limits his alcohol and soda intake.

2.1 Make a meal plan for an entire week. This plan should feed a typical 18- to 22-year-old college student, gender of your choice. Include all the meals, snacks, beverages, etc., that they would typically consume for an entire week Monday–Sunday.

	MON	TUE	WED	THU	FRI	SAT	SUN
Breakfast							
Lunch							
Dinner							
Snacks							

2.2 Determine the calorie content of the week of food and beverage for your typical college student. Use calorie calculators online or on food labels.

Jack knows that calories are just a unit of measure for energy. Calories are neither good nor bad, just like no food is good or bad. Knowing how much of each food to have is the key to keeping a calorie balance. Though Jack has never been really skinny, he also has never been obese. He keeps fairly fit by keeping his diet in check because he really hates to work out. Some people may think he needs to lose a few pounds, but he is happy and his wife Diane loves him just the way he is. In order to keep his weight in check without exercise, Jack pays attention to "calories in, calories out."

2.3 Jack is a 35-year-old male who is 6'2" tall and weighs 205 pounds. He is an architect who goes to some job sites but mostly stays in the office and attends meetings. He spends the weekends playing in the park with his kids and sometimes will meet up with his buddies to play a game of flag football or basketball, depending on the season. Calculate his BMI and then determine his basal metabolism, physical activity, and dietary thermogenesis calorie needs. That is, determine how many calories he needs to eat each day to maintain his weight.

2.4 Compare Jack's calorie needs to that of your typical college student. Would Jack lose or gain weight if he ate like your college student? Explain your reasoning.

2.5 Diane is a 35-year-old female who is 5'7" tall and weighs 195 pounds immediately after the birth of their twins. She will be staying home with the babies for a couple of years, taking care of all household chores and the twins while Jack is at work. (Jack helps with the girls once he gets home from work.) Calculate her BMI and then determine her basal metabolism, physical activity, and dietary thermogenesis calorie needs. That is, determine how many calories she needs to eat each day to maintain her weight.

Having given birth to twins, Jack's wife Diane is feeling a little down about how long it is taking her to get back into shape. She sees all the movie stars give birth one week and are back down to a size 0 two weeks later and just doesn't know how they can do it. (What she doesn't realize is that they have personal trainers working with them through the pregnancy and after giving birth. If Diane's only job was to work out and had a professional teaching her, she would lose the weight more quickly too. Too bad that isn't real life.) She asks Jack for his help to lose the last 10 pounds. Since she wants to lose only about a pound a week, she knows she needs to make a calorie deficit of 500 calories each day, or 3,500 calories for the week. Together, they adjust her meal plan so she has a 300-calorie deficit each day, but she realizes that she will need to work out for the other 200 calories each day.

2.6 Diane has been busy at home with twin babies in the house. Find exercises she can do either with the family or in what little spare time she has to burn the necessary 1,400 calories each week. Cleaning the house is something she already does; she needs exercises above and beyond those activities. At this point, she cannot afford a gym membership (like she has time to go anyway) so the activities need to be doable without fancy equipment. She would like a variety of options, not just running each day. Give her a two-week schedule with a different activity each day.

As the twins grow from newborns to toddlers to active youngsters, their dietary needs are changing. With the good habits Jack and Diane have developed, they feel confident in being good role models of healthy eating and an active lifestyle.

5.

NUTRITION

Nutrition is the study of foods, their nutrients and other chemical constituents, and the effects of food constituents on health. Food is a basic need of humans. We need enough food to live and the right assortment of foods for optimal health. Food security is a state where you have access at all times to a sufficient supply of safe, nutritious foods. Even in the United States we do not always experience food security. Food security also involves being able to acquire the food in socially acceptable ways. Food insecurity is the opposite; food insecurity existed in about 12% of U.S. households in 2002 and rose to 14.9% in 2011.

Foods provide energy (calories), nutrients, and other substances needed for growth and health. We need food calories to go about our day, nutrients to feed our growth and health, and other substances. We do eat for other reasons, though. Nutrients are chemical substances present in food that are used in the body. Essentially everything that is in our body was once a nutrient in food we consumed. Health problems related to nutrition originate within our cells. Poor nutrition can result from both inadequate and excessive levels of nutrient intake.

Humans have adaptive mechanisms for dealing with fluctuations in dealing with nutrient intake. Malnutrition can result from poor diets and disease states, genetic factors, or a combination of these causes. Malnutrition means "poor" nutrition and results from both inadequate and excessive availability of calories and nutrients in the body. Vitamin A toxicity, obesity, vitamin C deficiency (scurvy), and underweight are example of malnutrition. This can result from poor diets and also from diseases that interfere with the body's ability to use the nutrients consumed. Diarrhea, alcoholism, cancer, bleeding ulcers, and HIV/AIDS, for example, may be responsible for the development of malnutrition. There are also genetic factors. You may inherit high cholesterol even though you eat what a cardiologist would consider to be a heart-healthy diet. You may be pre-diabetic

due to genes passed through generations. These people, with genetic predispositions, are at a higher risk of becoming inadequately nourished than are others.

Poor nutrition can influence the development of certain chronic diseases. Faulty diets play important roles in development of heart diseases, hypertension, cancer, osteoporosis, and other chronic diseases. Heart disease has been shown to be linked to diets high in animal (saturated) fat and cholesterol intakes, and low intakes of certain vitamins and minerals, and fruits and vegetables. Excessive body fat is also a contributor to heart disease. Cancer has been linked to diets with low consumption of fruits, vegetables, and fiber along with excessive body fat and alcohol intake. Adult-onset diabetes has been linked to excessive body fat, low fruit and vegetable intake, and high saturated fat intake. Cirrhosis of the liver may be caused by excessive alcohol consumption along with a poor overall diet. Hypertension, also known as high blood pressure, can be affected by diets high in sodium and alcohol along with excessive levels of body fat. Iron-deficiency anemia is frequently caused by low iron intake. Tooth decay and gum disease is frequently caused by diets with excessive sugar consumption and inadequate fluoride intake. Osteoporosis comes from diets with an inadequate amount of calcium and vitamin D along with other factors.

Many fads will tell you to stay away from "bad" foods and eat only "good" foods. Nearly 80% of adults believe that a food can be good or bad. Spoiled and rotten food is bad, but otherwise bad foods do not exist. Foods can be healthy or unhealthy, not good or bad. Nearly all foods can be used wisely to contribute to a healthy diet. The key to a healthy diet comes down to adequacy of quality foods, variety, and balance.

NUTRITION LABELS

Nutrition labels are required for processed foods and supplements in the United States, but fresh fruit and vegetables along with raw meat are not required to have labels. Nutrition labeling rules allow health claims to be made on the packages of certain foods and products, but the claims must be truthful and adhere to FDA standards. Due to the size of the packaging, nutrition labels are not able to contain all the information people need to make healthy decisions about what to eat.

Nutrition labels are required to have certain components. The figures show a nutrition label for macaroni and cheese broken into several parts. The first thing to look for is the serving size.

(2A)

Figure 2A: Nutrition label serving size.

This particular product has a serving size of 1 cup but 2 servings are contained in the package. This is important, as the calculations of the calories and nutrients you are consuming are based on one serving only. If you were to eat the entire contents of this package, you would be doubling the amounts listed on the nutrition label. The next thing to check is the calories.

Amount Per Serving	
Calories 250	Calories from Fat 110

(2B)

Figure 2B: Nutrition label calorie content.

A calorie is a measure of how much energy you can get from a particular food. Notice that the calorie count is per serving, so if you eat half the contents of this package, you will consume 250 calories, but if you eat the entire contents, you are consuming 500 calories. In general, a food with 40 calories is very low, 100 calories is a perfect snack as it is considered moderate, 400 calories or more is high for one food item. Next, we move down to the nutrients.

	% Daily Value*
Total Fat 12g	18%
Saturated Fat 3g	15%
Trans Fat 1.5g	
Cholesterol 30mg	10%
Sodium 470mg	20%
Total Carbohydrate 31g	10%

(2C)

Figure 2C: Nutrition label nutrient content.

The first three listed are total fat, cholesterol, and sodium. You want to limit these as much as possible. The percentages listed at the right are based on a 2,000-calorie diet; that is, a diet in which you take in 2,000 calories per day. The carbohydrates are not as big of an issue as some fad diets may lead you to believe. You do want to limit your sugars, but dietary fibers also provide carbohydrates in your diet and these are something you should try to increase.

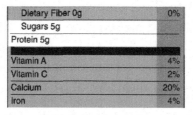

Dietary Fiber 0g	0%
Sugars 5g	
Protein 5g	
Vitamin A	4%
Vitamin C	2%
Calcium	20%
Iron	4%

(2D)

Figure 2D: Nutrition label nutrient content.

Along with fiber, the vitamins and minerals that come next you will want to make sure you are taking in an adequate amount daily. The percentages do not have to be listed, but when they are, they always look the same. A general guide is that 5% or less is low, whereas 20% or more is considered high.

* Percent Daily Values are based on a 2,000 calorie diet. Your Daily Values may be higher or lower depending on your calorie needs:			
	Calories:	2,000	2,500
Total Fat	Less than	65g	80g
Sat Fat	Less than	20g	25g
Cholesterol	Less than	300mg	300mg
Sodium	Less than	2,400mg	2,400mg
Total Carbohydrate		300g	375g
Dietary Fiber		25g	30g

(2E)

Figure 2E: Nutrition label percent daily values.

Many common terms are included on food and supplement labels. Calorie-free foods, such as Crystal Light, do not necessarily contain zero calories. If a product claims that it is a "light" or "lite" version of the original product, then it is required to contain 1/3 the calories or 1/2 the fat per serving of the original or a similar product. If a product claims that it is somehow "reduced," then it must contain at least 25% less per serving than the reference food.

 Everyday Health with Jillian Michaels—Understanding nutrition labels.

CALORIES

Some diets claim you should increase your protein and cut the carbohydrates. Other fads claim that you should limit your protein and go crazy on breads. How can they both be right? They can't. Gram for gram, carbohydrates provide the body with the same amount of energy that protein does at 4 calories per gram for each. Fat provides 9 calories per gram and alcohol provides 7 calories per gram. In terms of nutrient density, you will get more benefits from a gram of protein or carbohydrates than you would from a gram of fat. This is not to say that you should avoid fats altogether. Healthy fats are a part of a varied and healthy diet. There are also many discussions about whether butter or margarine are worse for you. Both butter and margarine have the same amount of fat, so they both provide the same number of calories (about 35 calories per teaspoon).

A calorie is a unit of measure used to express the amount of energy produced by foods in the form of heat. Everyone needs a daily intake of calories to provide our bodies with the energy to keep our heart pumping, work our lungs, and digest foods. The energy we need breaks down into three main categories: for basal metabolism, for physical activity, and for dietary thermogenesis.

The calories necessary for basal metabolism provide energy to support body processes such as growth, health, tissue repair, and maintenance, as well as many other functions. Basal metabolism includes energy the body expends for breathing, the pumping of the heart, maintenance of body temperature, and other life-sustaining, ongoing functions. The most variance in calorie needs comes from physical activity. An inactive person is sitting most of the day with less than two hours of moving about slowly or standing. Most office workers are considered inactive during the work week. A person with average physical activity is sitting still most of the day but walking or standing for two to four hours with no strenuous activity. A professor would be an example of a person with average physical activity, unless they do most of their lectures sitting down, then they would also be inactive. A physically active individual is active four or more hours each day with little standing or sitting, and the inclusion of physically strenuous activities. A construction worker and an active soldier are examples of physically active individuals. The last category, dietary thermogenesis, refers to the calorie (energy) needs for digesting foods, absorbing and utilizing nutrients, and transporting nutrients into cells.

BASAL METABOLISM

Basal metabolism makes up the bulk of most people's needs for calories with a range of 60% to 80% of the total needs. A quick estimate for basal metabolism calorie requirements is

- For men: Multiply body weight in pounds by 11
- For women: Multiply body weight in pounds by 10

Example 1: Dave weighs 180 pounds and Connie weighs 145 pounds. Find the calorie needs for each of them just for basal metabolism.

Solution 1: Dave is a man so his basal metabolism calorie needs are 180(11) = 1,980 calories. Connie is a woman so her basal metabolism calorie needs are 145(10) = 1,450 calories.

PHYSICAL ACTIVITY

Physical activity accounts for the second highest number of calories we expend. The three basic lifestyles mentioned previously have different percentages necessary for calorie calculation.

- Inactive: Multiply basal metabolism needs by 30%
- Average: Multiply basal metabolism needs by 50%
- Active: Multiply basal metabolism needs by 75%

Example 2: Neil is a road contractor and is very active each day. Gennie stays home with the children so leads an average lifestyle. If Neil's basal metabolism is 2,140 calories and Gennie's basal metabolism is 1,310 calories each day, find the number of calories they each need for physical activity.

Solution 2: Since Neil is active, we multiply his basal metabolism by 75%: 2,140(0.75) = 1,605 calories for physical activity. Using the average calculation, we multiply Gennie's basal metabolism by 50%: 1,310(0.50) = 655 calories needed for physical activity.

Dietary Thermogenesis

Calories expended for dietary thermogenesis are estimated at 10% of the sum of basal metabolic and usual physical activity.

Example 3: Chad has calculated his basal metabolism to require 1,870 calories and his physical activity to require 1,402 calories. Jodi has calculated her basal metabolism to require 1,300 calories and her physical activity to require 650 calories. Find the required calories for dietary thermogenesis for Chad and Jodi.

Solution 3: Chad will find 10% of the sum of basal metabolism and physical activity so he will calculate $(1,870 + 1,402) \times 0.10 = (3,272)(0.10) = 327.2$ calories for dietary thermogenesis. Similarly, Jodi calculates $(1,300 + 650) \times 0.10 = (1,950)(0.10) = 195$ calories for dietary thermogenesis.

To determine your total calorie needs each day, you would then add the values from basal metabolism, physical activity, and dietary thermogenesis. There are online calculators that will do this for you, but it is always good to know how it is done and to be able to do it for yourself.

Energy Balance

If you are maintaining your weight, the number of calories that you need each day is equal to the number of calories that you are consuming; this is known as energy balance. Energy balance can be affected by smoking, lean muscle mass, and genetic makeup. In general, smokers burn slightly more calories than people who do not smoke. This is not a good enough reason to take up smoking, however, as the calorie count is about 100 per year and the negative health effects far outweigh the small benefit. Muscular individuals need more energy to maintain body weight than do people of the same weight with less muscle. Muscle needs more calories than fat and muscle will burn more calories than fat. The genetic factor is very slight, but it is significant enough to mention in this discussion.

Weight Status

Underweight, normal weight, and overweight are generally defined based on weight-for-height charts frequently found in doctor's offices. The term "obese" is based primarily on body fat content. An individual may be labeled overweight based on weight-for-height calculations, but that does not necessarily make them obese. To calculate weight-for-height, we again have to divide into gender differences, using the Hamwi Formulas:

- For women: begin with 5 feet of height equals 100 pounds then add 5 pounds for each additional inch of height. The standard weight is plus or minus 10% of this amount.

- For men: begin with 5 feet equals 106 pounds then add 6 pounds for each additional inch of height. The standard weight is plus or minus 10% of this amount.

Example 4: Donald is 6'1" and Sherri is 5'3"; find their target weight-for-height ranges.

Solution 4: Using the calculation for men, Donald is 5 feet (106 pounds) plus 13 inches (13 × 6 = 78 pounds). Donald should be 184 pounds with a range of 184 − 10%(184) = 165.6 pounds to 184 + 10%(184) = 202.4 pounds. For Sherri, she is 5 feet (100 pounds) plus 3 inches (3 × 5 = 15 pounds). Sherri should be 115 pounds with a range of 115 − 10%(115) = 103.5 pounds to 115 + 10%(115) = 126.5 pounds.

Body Mass Index (BMI)

Body mass index is a calculation that is the same for men and women, for inactive and active. The formula is:

$$BMI = \frac{weight(lb)}{\left[height(inches)\right]^2} \times 703$$

Example 5: Jeff is 6'4" and weighs 160 pounds. Julie is 5'6" and weighs 130 pounds. Calculate each one's BMI.

Solution 5: First, we convert the standard height into a height given purely in inches. Jeff is 6'4" so is 6 × 12 + 4 = 76 inches tall. Julie is 5'6" tall so is 5 × 12 + 6 = 66 inches tall. For Jeff, $BMI = \frac{160}{[76]^2} \times 703 = 19.47$ and for Julie, $BMI = \frac{130}{[66]^2} \times 703 = 20.98$.

In the previous example, both Jeff and Julie have a weight status termed normal. There are five categories of weight status for BMI calculations:

BMI	Weight Status
Below 18.5	Underweight
18.5–24.9	Normal
25.0–29.9	Overweight
30.0–34.9	Obese
35.0 and Above	Morbidly Obese

Figure 2F: BMI and weight status.

The fact is everyone needs some body fat. Our bodies use it on a daily basis. Men require 3% to 5% and women require 10% to 12%; if you have less than that you would no longer be considered healthy. Too much body fat is also an issue, as 20% or more for men and 25% or more for women is considered obese.

WAIST-TO-HIP RATIO

The waist-to-hip ratio is another measure to determine if you are storing necessary body fat in a healthy way. In order to calculate, for both men and women, locate your waist. This is the smallest circumference below the last rib of the rib cage and above the navel. Next, locate your hips. This should be measured across the largest circumference of the buttocks. Once you measure your waist and hips, divide your waist circumference by your hip circumference, using consistent units (i.e., inches). For women, the ratio should be less than 0.80 and for men the ratio should be less than 0.95.

CALORIES IN, CALORIES OUT

All this information can seem overwhelming at times but it all comes down to calories in and calories out. Theoretically, about 3,500 calories is equal to one pound of weight. To lose one pound per week, you need to consume 3,500 fewer calories that week than you use. You can break this into a calorie deficit of 500 calories per day in which you burn more than you consume. Similarly, if you are trying to gain weight, you would want to use the same calculations, making sure that your food choices are healthy in order to add weight as muscle and not as fat. When cutting calories, our bodies will start expending calories at a slightly lower rate. This is known as survival instinct that is ingrained in our bodies; they fear starvation. This is why you should not make drastic changes to your calorie intake; make all changes slowly to allow your body time to adjust to the new levels. Changes in increments of 10% are best for keeping your body working correctly. Diet and exercise CAN work by itself. You do not need supplements, diet foods shipped to you, or a group of people to tell you what to do after you pay them money. Keep track of the foods you eat, know how many calories you burn per day, and you can keep your weight where you want it. (Some diseases make this more challenging. This is not a statement against any system, product, or plan; it is also not to replace a doctor's information and suggestions.)

 Ask Jillian Calories in vs Calories out.

It is not true that you can achieve any body weight or shape you desire if you work hard enough. Some of us just do not have the genetic materials for six-pack abs. This does not make us any less healthy. Half of the women in the United States wear size 14 to size 26, yet many clothing models and actresses are severely underweight. The best way to prevent obesity is to become more accepting of people of different sizes and shapes. Healthy is not in a clothes size or even a weight; healthy comes from a regular intake of healthy foods along with a lifestyle of healthy activities.

IN- AND OUT-OF-CLASS ACTIVITIES

1. Keep a food journal for one week. Include all beverages, food, and relative amounts. Once the week is complete, research to find the calorie counts for each day.
2. Use your personal food journals to determine if you should expect to lose weight, maintain weight, or gain weight based on your activity level.

CHAPTER THREE

PROBABILITY

6.

J ack was able to get his driver license three weeks after his 16th birthday. He thought that day would never come! Having taken a driver safety course, he didn't need to take a road test with an officer; he thought that would be way too stressful. He passed the written portion of the exam, missing only one answer; he was definitely prepared to hit the road.

3.1 While applying for his driver license, Jack is looking at his social security card. He notices that it has nine digits. If every nine-digit number were available for the Social Security Administration, how many social security cards could they issue?

3.2 Find the current estimated population of the United States. With this population, is it reasonable for each resident to have their own social security number? Do you think the United States will ever reach a situation when one social security number must be assigned to two people? What ramifications might this have?

3.3 The first social security numbers were issued in 1936. Find the population of the United States in 1936. What is the fewest number of digits a SSN could have had in order to assign one to each resident of the United States that year?

Jack has his license so now his dad has to add him to the insurance. The insurance company requires the license plate number and vehicle identification number (VIN) for each registered automobile. Jack starts paying attention to license plates now.

3.4 In California, one particular license plate design has a numeric digit followed by three letters followed by three more numeric digits. For this one plate design, how many license plates could be made in California?

3.5 In New York, one particular license plate has three letters followed by four numbers. For this one plate design, which state has more plate options, New York or California? Explain your findings.

3.6 In New Hampshire, one particular license plate has three digits followed by four digits. Will New Hampshire have more plates available for this single design or fewer than New York and California? Why do you think New Hampshire can use this system and not need the letter characters?

7.

One of the scholarships that Jack received as a freshman college student was the lottery scholarship. This gave him enough money each semester to cover the cost of books plus some spending money. He had never really given much thought to the lottery, even though he saw the billboards claiming he could be a millionaire, but he decides now that he is a college student with some spending money, he's going to look into it some. If he likes what he finds, maybe he'll buy a ticket and quit college!

Jack does a Google search to find some information on lotteries. He sees two types of lotteries that he is familiar with: scratch-off tickets and user-selected number games. He has seen Cash-3 in the convenience store so he reads more about it.

3.7 If Jack bets the numbers 4-2-7 on Cash-3, what it the probability that he wins? Will it make a difference if he bets straight as opposed to box? If so, what are the different probabilities?

3.8 If the winning straight ticket earns $5,000 and the winning box ticket wins $2,000 for Cash-3, what is the expected value for each type of bet if each ticket costs $1 and you either win or you lose?

3.9 If Jack bets the numbers 3-3-6-9 on Pick-4, what is the probability that he wins? Will it make a difference if he bets straight as opposed to box? If so, what are the different probabilities? (Note the double three and how this may change the box betting outcomes.)

3.10 If the winning straight ticket earns $7,500 and the winning box ticket earns $3,500 for Pick-4, what is the expected value of each type of bet if each ticket is $1 and you either win or you lose?

The lottery games Jack has seen the most advertisements for are the Powerball and Mega Millions. He remembers hearing about people winning millions in the lottery and then being poor just a couple years later; he can't imagine spending all that money so soon. Looking into the rules and pay-outs of Powerball and Mega Millions, he starts to think this lottery stuff is a lot more complicated than it seems.

3.11 Jack decides to spend $2 on a Powerball quick-pick ticket where the computer picks his numbers for him. What is the probability that Jack will win the jackpot? What is the probability that Jack wins anything at all?

3.12 Jack decides to spend $1 on a Mega Millions quick-pick ticket. What is the probability that Jack will win the jackpot? What is the probability that Jack wins anything at all?

3.13 Jack starts to believe he shouldn't waste money on lottery tickets, so he wants to figure out when it is a good idea. For a $2 bet in Powerball, find the amount of the jackpot necessary for the expected value to be $0. That is, he doesn't win anything but he also doesn't lose money on average.

3.14 For a $1 bet in Mega Millions, find the amount of the jackpot necessary for the expected value to be $0.

8.

Jack decides that the lottery isn't for him. He appreciates the help of the scholarship, but he realizes that is not the way to make a quick buck. Sitting around thinking of other get-rich-quick schemes, Jack hears a commercial on the radio for the local horse track. He has seen the movies where people place one little bet and win thousands. Maybe betting on the horses is his way to making millions! Before he goes to the track, he does a little research on the betting principles and pay-offs.

3.15 Jack's favorite color is red so he bets $5 on the horse wearing the red number to win. If the odds of this horse winning are 5-2, how much should Jack expect to collect if this horse wins the race?

3.16 If Jack decided to bet $5 on the horse with the funniest name instead, how much should he expect to collect if the odds are 10-1?

Jack realizes that he doesn't have enough money to place big bets in order to win big pay-offs. He's disappointed, but since he's allergic to horses anyway, he feels avoiding the track is probably a good idea. Back in his dorm room studying for his math test, Jack hears that his favorite band is playing a free concert at a local casino. It occurs to him that he could make a killing at a casino! He just needs to understand a few basic games, start with some seed money, and then use his intelligence to make his money grow. Having a hard time focusing on the math now (because of the music and money) he decides to research casino games. Before he can understand some of the basic games on the casino floor, he realizes that he needs to learn the math involved. Math is everywhere! Oh well, he thinks. At least this math can make him money.

Jack arrives at the casino an hour or so before the Somebody's Weird concert so he can try his first game. He has $60 to spend, $20 at each of three games. He plans to place only bets of $5.

3.17 Jack decides to start at the Roulette table. His first bet is on 1 – 18. If the ball stops with a favorable outcome, how much will he win?

3.18 Jack takes those winnings and puts them in his pocket. He takes his next $5 and bets on the second column. Luck must be on his side today: the ball lands on 17 and he wins! How much money does Jack win on this spin of the wheel?

3.19 Now that Jack is all excited about winning two spins in a row, he throws caution to the wind and puts two $5 bets down at the same time. He bets a corner of 20-21-23-24 and also the second twelve. If the ball lands on the number 24, his lucky number, how much money will Jack win? What are Jack's total winnings from Roulette?

That was exhilarating! Jack is so excited about winning that he goes into the concert with a smile that never ends. He can hardly believe he gets to see his favorite band for free and he won money all in the same day. After the amazing show, Jack returns to the casino floor to try his luck at the next game. The card suits will be represented as diamonds = (♦), clubs = (♣), hearts = (♥), and spades = (♠).

3.20 Keeping to his plan, Jack now moves to the blackjack table. He had $20 set aside for use at the blackjack tables, too, so he isn't touching any of his winnings. Jack puts down his first $5 bet and is given the two cards 5(♥) and 10(♦). The dealer, the only other "player" at the table, shows an 8(♣). Should Jack hold or hit? Why or why not? Jack decides to hit and gets a 4(♥). Should Jack hold or hit? Why or why not? Jack decides to hold. The dealer's down card is a K(♥). Jack wins!!! How much does he win?

3.21 Jack lays down his next $5 bet. This time, after the dealer has finished, Jack sees that he has a J(♠) and an A(♥) and the dealer has a 2(♦)! This is known as a blackjack, though not in the strictest sense, and is the point of the game; how much does Jack immediately win?

3.22 On his third turn, Jack once again lays down his $5 bet. This time he is dealt an 8(♠) and a 3(♥). Added together, this becomes an eleven. Jack glances at the dealer's up card and sees a 6(♣). Jack decides to "double down" and so places his last $5 on the table. His next card is a Q(♥)! Watching as the dealer draws to a 20, Jack learns he wins!!! How much does Jack win on this bet? How much did Jack win in total playing blackjack?

Having brought only $60 to bet in the casino, Jack is pretty excited to try his third game. He is having a difficult time deciding between stud poker and Texas Hold 'Em. He watches both games for a while and then decides maybe he'll just play Red Dog instead.

3.23 Jack watches Red Dog for a little while and decides that it is a straightforward game and he is ready to play. He places his initial $5 bet and the cards dealt are a 3(♥) and 7(♠). As this is his first time playing, he decides not to raise the bet. The next card dealt is a 5(♥). How much does Jack win?

3.24 For the next deal, Jack once again places a $5 bet. This time the cards show 8(♦) and J(♣). He sees the other players raising, so he raises his $5 bet. If the next card is a 9 or 10, how much will Jack win? Unfortunately the card was an A(♦) and so Jack lost his money.

Having only $5 remaining, Jack places his bet one last time. This time the cards are 7(♣) and Q(♦). He is tempted to use his winnings in order to raise but decides not to give in to temptation. The card dealt is a 2(♠). Jack loses his last bet. As he walks back to his car, Jack is thinking about his winning and losing. He realized that he could quickly get addicted to the winning but losing always hurts.

3.25 Jack is sitting in his car and decides to count his winnings before leaving the parking lot. How much did Jack win at the casino on this trip?

6.

PROBABILITY BASICS

In order to understand the basics of probability, we must have a few basic definitions.

Definition 1: An experiment is an occurrence with a result or outcome that is uncertain before the experiment takes place. The set of all possible outcomes is called the sample space for the experiment.

An experiment could consist of something as simple as flipping a coin or choosing a card from a deck. It does not have to be an experiment in the strictly scientific sense. In this text, experiments will be Powerball or Mega Millions lottery draws, horses running a race, or a spin of the Roulette wheel.

Definition 2: Given a sample space S, an event E is a subset of S. The outcomes in E are called favorable outcome. We say that E occurs in a particular experiment if the outcome of that experiment is one of the elements of E—that is, if the outcome of the experiment is favorable.

At the start of every NFL game, the team captains meet with the head referee at centerfield for a coin toss. If the Seattle Seahawks call "heads" during the flip, the favorable outcome for the Seahawks is the coin landing heads facing upward. In this case, we say the sample space S consists of heads and tails with the event space for the Seahawks as heads. Note that the favorable outcome for their opponent is tails; we call this the complement.

Definition 3: The complement of an event E is the set of outcomes not in E. Thus, the complement of E represents the event that E does not occur.

To calculate a probability, we have two main methods. The first method is what we observe; this is called experimental, or empirical, probability. With experimental probability, we conduct the probability experiment numerous times, recording the results as we go. Perhaps we decide to determine the probability of flipping a coin and having it land heads up. In order to conduct this experiment, we choose a coin and flip it 50 times, recording the results after each toss. At the end of the experiment if we discover that heads landed up 37 times, our probability of flipping heads has an empirical probability of 37/50. The other way to compute a probability is using mathematics to determine what should happen. When tossing a coin there are two results possible: heads or tails (neglecting the possibility the coin never comes down or it somehow lands on its edge). Out of these two results, one is heads. Thus our theoretical probability of flipping a coin and heads landing up would be 1/2—one favorable result divided by 2 total results. We can also think of this as the number of favorable outcomes divided by the total number of outcomes possible.

When dealing with probabilities, we must keep in mind three "rules." These rules are known as a probability distribution and are one of the most important things to keep in mind when calculating and working with probabilities.

Properties of Probability Distributions:

1. The probability of every element in the sample space must be between 0 and 1, inclusive. That is, the probability that any result happens must be 0 (never going to occur) through 1 (certain to happen). Probabilities are frequently written as fractions, but decimal representations are also allowed.
2. If we add the probabilities of every element in the sample space, we must get 1. This tells us that the sample space must contain every possible outcome of the experiment. The sum must be 1, as this represents a 100% chance that something from the sample space occurs. Note that you cannot have a 120% = 1.2 probability that a particular result occurs.
3. If you have an event E that is a subset of your sample space S, then we can add the probabilities of all possible results in E to get the probability of E itself. For example, if we want to win something (anything!) on our Powerball ticket, the probability of winning would be to add the probability of each possible winning combination.

Definition 4: The Law of Large Numbers states that in order for an empirical probability to be close to the value of the theoretical probability, you must conduct the experiment a large number of times.

In a previous example I flipped a coin 50 times and it landed heads up 37 times. Our experimental probability of 37/50 is not very close to the theoretical probability of 1/2. However, if we were to conduct our experiment 500 times, we would find that our experimental probability would be closer to a result of 1/2 = 0.5 = 50%.

ADDITION AND MULTIPLICATION PRINCIPLES

When dining at a restaurant you have many options available to you. Some of these options require you to use the addition principle of counting and others require you to use the multiplication principle of counting. The addition principle states that when presented with *n* different alternatives, each with a certain number of outcomes, then you simply add the outcomes to arrive at the total number of alternatives. For example, you go to a restaurant that has a lunch menu with three different soups, four types of salads, four types of burgers, five types of sandwiches, and three fish baskets. From this menu you have $3 + 4 + 4 + 5 + 3 = 19$ different selections for lunch.

The multiplication principle is similar with one main difference: you use the multiplication principle if you have a step-by-step process. Suppose you were visiting a restaurant offering a three-course meal. They have four appetizers to choose from, five main courses, and two desserts. Using the multiplication principle we find that there are $4 \times 5 \times 2 = 40$ different three-course dinners possible. As you can see, the *multiplication principle* is used when there is a *sequence* of choices, whereas the *addition principle* is used when you just have several *alternatives*.

We can use these two principles together for decision making. You decide to go to lunch at your favorite restaurant that offers a 2-for-$4 lunch option. You can pick a soup or salad and then match it with a sandwich or burger. If the restaurant offers three different soups, four different salads, four different burgers, and six types of sandwiches, how many days could you go to the restaurant for a 2-for-$4 deal and have a different meal each day? We will use the addition principle to pair $3 + 4$ soups or salads with $4 + 6$ different sandwiches or burgers. As we make one choice and then the other, we finish with the multiplication principle $(3 + 4) \times (4 + 6) = 12 \times 24 = 288$. We could go to the restaurant for 288 straight days and never have the same meal! This process, using both the addition and multiplication principles, is called a decision algorithm. It is important to realize when you use alternatives (addition) and steps (multiplication).

Other than just food, we see the use of the multiplication principle every day. When there were fewer cars on the roads, license plates started at 1 and went to a number as high as necessary for that particular county or state. With over an estimated 254 million cars now registered in the United States alone, a better system had to be created. Most standard license plates allow for six characters. Some states, such as New Mexico, have three alphabet characters followed by three numeric digits (or vice versa depending on the plate design). The use of just numbers for a 6-digit license plate would allow for $10 \times 10 \times 10 \times 10 \times 10 \times 10 = 10^6 = 1{,}000{,}000$ different license plate combinations. By allowing the use of letter characters as well, the state of New Mexico has $26^3 \times 10^3 = 17{,}576{,}000$ different license plates available for each design. (Note that not every three-letter combination of letters is allowed, as profanity is banned from automobile license plates.) One simple change from 10 digits to 26 letters allowed for more automobiles to be registered in the state, using just one design of license plate, than people living there (estimated at 2 million residents in 2012). Consider also a combination lock as a sequence of steps. To try to guess a three-number combination on a safe with values 00–39 would take up to 64,000 guesses!

7.

LOTTERIES

Lotteries are popular games today, but they have been around for centuries. Since the beginning, lotteries have been a way for governments to raise money for public works. One of the first lotteries may have occurred during the Han Dynasty in China (between 205 and 187 B.C.) and is thought to have helped raise money for the construction of the Great Wall of China. This first type of lottery game was Keno, a game that is still played today.

Ancient texts from civilizations including Ancient China, the Celtic Era, and even Ancient Greece as mentioned in Homer's *The Iliad*, have all referenced lotteries and "drawing lots." Though the games started out as entertainment, Augustus Caesar used lotteries to raise money for city repairs in Rome.

There are four main types of lottery games today: scratch-off tickets, ready-printed numbers, user-selected number games, and video-based games. Scratch-off tickets are one of the quickest lottery games in that you find out if you are a winner as you scratch off the metallic portion of the ticket. Some states have crossword scratchers, scratch tickets that resemble slot machines, holiday tickets, and many more. The state of New Mexico has over 50 different scratch-off ticket types. Ready-printed numbers are used frequently in less-developed countries where complex communication networks are less available. They are also used as raffle tickets during high school sporting events, events with door prizes, and other fundraising type activities. The United States, Canada, and the UK play user-selected number games extensively. These are the games we will focus on.

User-selected games come in several types: one-number, multi-number, multi-number with two sets of numbers, and keno. The one-number is most commonly a three- or four-digit number. The three-digit games are called Cash-3, Pick-3, Play-3, and so on. In this game you pick any three-digit number from 000 to 999. Similarly, in a four-digit number game, you choose any four digit number from 0000 to 9999. In the multi-number game, a user selects 4, 5, or

6 numbers from 1 to *n*, where *n* is a predetermined ending number. The classic lotto was a multi-number game in which you chose six numbers from 1 to 49. Some lotteries pay for getting as few as two numbers correct. The "jackpot" is when you get all the numbers correct. Though your numbers will be printed on the ticket from least to greatest, the order in which the numbers appear is irrelevant to winning. Multi-number with two sets of numbers is as the category suggests. The most common of these games are Powerball and Mega Millions. In Keno, you select 20 numbers (usually from 1 to 80) and if 10 of them come up, you win. This is a game frequently seen in restaurants all over Nevada as well as many truck stops throughout the country.

 What to do if you win the lottery.

Cash-3

In a game of Cash-3, or any other name it may go by, you pick a three-digit number from 000 to 999. Suppose we chose the number 123. We now have the opportunity to pick one of two types of bets. A straight bet indicates that you think it will be an exact match. That is, you think the number 123 will win. A box bet indicates that you think it will be those digits, but a possibly different order. You can win with 123, 132, 213, 231, 312, or 321. Other bets you may place include combination, front pair, back pair, and split. However, these depend on the state in which this lottery is held. We will stick to the basic straight and box bets for now.

Question 1: How many numbers are there from 000 to 999?

Question 2: If you choose a straight bet, how many different winning numbers are there?

The answer to question one is a simple issue of counting. We know that from 1 to 999 is 999 numbers. If we add in 000 we now have 1,000 numbers from 000 to 999. We could also use the multiplication principle to answer this question. There are ten digits available for each number, three numbers being asked for, so there are $10^3 = 1,000$ different three-digit numbers. The answer to the second question is easier, but slightly trickier. When betting a straight bet, you are claiming that you have the one correct number. That is, the answer to question two is 1. There is only one winning number for a Cash-3 straight bet.

Question 3: What is the probability you win a Cash-3 straight bet?

The number of favorable outcomes is the answer to question two and the total number of possible outcomes is the answer to question one. That is, the probability that you win a Cash-3 straight bet is 1/1000. In a box bet, we have the same 1,000 numbers from 000 to 999. For simplicity, we will use three different digits; our pick is 456.

Question 4: How many different ways could we win a box bet in a Cash-3 game?

In order to answer this, we need a concept from probability called a permutation. A permutation is an ordered list of items. In this case, we want to put the digits 4, 5, and 6 in as many different orders as possible. For a three-digit number containing only these digits, we have three choices for the first number, followed by two choices for the second number (as one has already been placed in front), and finally only one digit available for the last number. If we multiply $3 \times 2 \times 1$, we will get 6 possible outcomes when we bet a box for a Cash-3 game.

Question 5: What is the probability that you win a box bet in a Cash-3 game?

Again we will divide favorable outcomes by total outcomes to get $\dfrac{6}{1000} = \dfrac{3}{500}$ as our probability. Notice that the box bet has a greater likelihood of occurring, as there are six chances as opposed to the one for a straight bet. Keeping this in mind, which do you think pays better when you win? Why?

POWERBALL

Powerball's first drawing, using that name, was April 22, 1992. Powerball was the first game of its type to use two containers for the ping pong–style balls with numbers. This allowed for a greater mix with many more prize levels, high jackpot odds, and low overall odds of winning. Powerball currently consists of a container holding white balls numbered 1 through 59, and another container holding red balls numbered 1 through 35. Out of the white balls, five are drawn with order not important. Out of the red balls, only one is drawn and this is called the Powerball. It costs $2 to play one set of numbers for Powerball, a price that increased from $1 on January 15, 2012. There is another option to add a Power Play option for an extra one dollar, but that is beyond the scope of this text. With the two container options, there are now nine ways that you could win with one selection of numbers:

Matches	Prize	Odds of Winning
Powerball only	$4	1 in 55.41
1 number plus PB	$4	1 in 110.81
2 numbers plus PB	$7	1 in 706.43
3 numbers; no PB	$7	1 in 360.14
3 numbers plus PB	$100	1 in 12,244.83
4 numbers; no PB	$100	1 in 19,087.53
5 numbers plus PB	$10,000	1 in 648,975.96
5 numbers; no PB	$1,000,000	1 in 5,153,632.66
5 numbers plus PB	Jackpot!	1 in 175,223,510

Figure 3A: Powerball odds and prizes.

These odds are extremely difficult to calculate, as they take into consideration many outcomes. A more complete treatment of probability is required. We will accept the lottery's numbers for the purposes of our discussion. Although given as odds, we will treat these as probabilities in fraction form for ease of calculation. That is, 1 in 55.41 has a probability of $\frac{1}{55.41}$.

MEGA MILLIONS

The Mega Millions game is very similar to the Powerball lottery game and was created due to the increased interest in Powerball. Sales for the first Mega Millions jackpot began on May 15, 2002 with the first drawing May 17, 2002. In the beginning, only nine states participated in the Mega Millions lottery. By 2013, all American lotteries participate in Mega Millions, including the Virgin Islands Lottery.

Keeping the cost at $1 per ticket does make for smaller starting jackpots in the Mega Millions game as opposed to the Powerball lottery; however, the jackpots can grow as large as, or even larger than, the Powerball during some periods of time. Mega Millions also offers the extra Mega-plier for $1 to multiply your winnings. We will again put this portion of the game aside for the sake of continuity. Mega Millions uses two containers to draw ping pong–style balls; one container holding white balls numbered 1 through 56 and the other container holding yellow balls numbered 1 through 46. As there are a different number of balls in each container than in Powerball, there are different odds of winning and different winning combinations. In California, all prizes are pari-mutuel, meaning pay-outs are based on sales and the number of winners (similar to horse racing). In all other states, prizes for 2nd through 9th are a pre-determined amount listed in Figure 3B. There are a total of nine ways to win in Mega Millions:

Match	Prize	Chances
5W +1Y	Jackpot	1 in 175,711,536
5W+0Y	$250,000	1 in 3,904,701
4W+1Y	$10,000	1 in 689,065
4W+0Y	$150	1 in 15,313
3W+1Y	$150	1 in 13,781
3W+0Y	$7	1 in 306
2W+1Y	$10	1 in 844
1W+1Y	$3	1 in 141
0W+1Y	$2	1 in 75

Figure 3B: Mega Millions chances and prizes.

PAY-OFF VERSUS EXPECTED VALUE

We play these games because of the numbers in the prize column, not because of the numbers in the chances or odds column. We buy a ticket dreaming of the jackpot, or even the $1,000,000 prize. It is the pay-off possibility that lures us in. Mathematically speaking, the pay-off is misleading. Expected value is a way of calculating how much you can expect to win, or lose, over the long range course of the game. In general, expected value is equal to the probability of winning multiplied by the pay-out for that win.

If we have three possible outcomes each with probability P_1, P_2, and P_3 and winnings of W_1, W_2, and W_3, respectively, you can find the expected value with the formula:

$$EV = P_1 W_1 + P_2 W_2 + P_3 W_3$$

That is, multiply the probability of an outcome by the winnings (or losings) for that outcome.

Example 1: Suppose I play a game with a die. If the number rolled is a 1 or 2, I win $7. If the number rolled is a 3 or 4, I win $3, and if the number rolled is a 5 or 6, I lose $4. Find the expected value.

Solution 1: For a game with a fair die, the probability of rolling any single number is 1/6. This means that the probability of rolling either a 1 or a 2 is 1/6 + 1/6 = 2/6 = 1/3. The same is true for rolling a 3 or 4, and a 5 or 6; each has probability 1/3. Notice that all the probabilities add to 1. We now have:

$$EV = \frac{1}{3}(7) + \frac{1}{3}(3) + \frac{1}{3}(-4)$$

$$EV = \frac{7}{3} + \frac{3}{3} - \frac{4}{3}$$

$$EV = \frac{6}{3} = \$2.$$

That is, when playing this game I can expect to win $2 per play on average. Note that it may take many plays of the game to hit that average. This is what is known as a favorable game because the expected value is a positive number.

Example 2: You have 10 marbles in a bag: 4 red, 2 blue, 3 orange, and 1 yellow marble. You pay $1 to play the game. You lose your bet if you draw a red marble. You win $10 if you draw the yellow marble, you win $5 if you draw a blue marble, and you win $2 if you draw an orange marble. What is the expected value of this game?

Solution 2: As there are 10 marbles in the bag, the probability of red is 4/10, the probability of blue is 2/10, the probability of orange is 3/10, and the probability of yellow is 1/10. Many times it is best to simplify our fractions; however, looking ahead we know we will need to add and subtract these fractions, so I will leave them over a common denominator. The expected value is

$$EV = \frac{4}{10}(-1) + \frac{2}{10}(5) + \frac{3}{10}(2) + \frac{1}{10}(10)$$

$$EV = \frac{-4}{10} + \frac{10}{10} + \frac{6}{10} + \frac{10}{10}$$

$$EV = \frac{22}{10} = \$2.20$$

Keeping in mind that it costs only $10 to play this game, we could expect to win $2.20 on average for each play. (Again, the number of plays must be large for this average to be met.)

8.

HORSE RACING

Archeological records indicate horse racing took place in ancient Greece, Babylon, Syria, and Egypt. Both chariot and mounted horse racing were events in the ancient Olympics by 649 B.C. As part of myth and legend, Norse mythology tells of a horse race between Odin and Hrungnir. Throughout the centuries, horses and horse racing have fascinated people. Thoroughbred racing was popular with royalty of British society and thereby earned the nickname the "Sport of Kings."

Nearly all the current thoroughbreds (the most commonly raced horse today) can trace back to one of three early sires: Darley Arabian, owned by Thomas Darley; Godolphin, owned by Lord Godolphin; or Byerly Turk, owned by Captain Robert Byerly. These three horses were taken to England where they were mated with racing mares, producing the stock we see in horse races even now.

In the United States, horse racing dates back to 1665 at the Newmarket course in Salisbury, New York. This track is now known as the Hempstead Plains of Long Island, New York. The first race was supervised by New York's colonial governor Richard Nicolls. In 1868, the American Stud Book was started and this marked the beginning of organized horse racing in the U.S. Twenty-two years later there were 314 horse racing tracks operating and in 1894 the American Jockey Club was formed.

Horse racing seemed to thrive in the United States from the mid-1600s through the end of the nineteenth century. The Puritans at the turn of the 20th century didn't like the betting aspect of the sports, so horse racing lost its shine. A change in betting style to pari-mutuel betting was introduced in 1908 and racing came back strong until the beginning of World War II. In order to gather renewed interest in the sport, the Triple Crown was introduced, bringing horse racing back to the American consciousness. The Triple Crown is a series of three major races: the Kentucky Derby, the Preakness Stakes, and the Belmont Stakes. If a single horse wins all three races in a given year, that horse is considered a Triple Crown winner. Only eleven horses have won the Triple Crown, with Sir Barton the first in 1919 and Affirmed the most recent in 1978.

What was this big change in betting all about? Pari-mutuel betting is a system in which all bets are placed together in a pool. Originally, this would happen at the track where the race occurred. With technological advancements of today, this now includes bets made from all over the world. Once all bets are made, the track (also known as the "house") takes their cut and determines the pay-out based on what remains. Due to the international business of horse racing, cut-off times are established prior to each race so that no more bets are possible. As the pay-out goes to everyone who bet a winner, the amount you win may change after you make your bet. If you are the first to bet on a particular horse, you may think that you will be earning some good money. However, as race time gets closer you see the possible winnings decreasing due to other bettors also wagering on the horse you chose. You must all share the winnings rather than you collecting them for yourself. The amount of pay-out depends on how much was bet in total, how many people bet on the particular horse, and on the odds that the horse will win the race.

 How odds work.

Odds do not represent the actual probability of an event happening; they represent the amount the house is willing to pay out for that event. Odds given in gambling are the odds against the activity happening. That is, if the odds are 2-1, the horse will lose two races for every one that it wins. Pay-outs for betting a 2-1 horse then mean that you win $2 for every $1 you bet, plus you get your original bet back. That is, betting $1 on a 2-1 horse will pay you $3, if it wins.

 Betting at the track.

There are many types of bets in horse racing, as the video described. This text is not intended to be a complete overview of horse racing nor is it intended to be a how-to guide. As with all gambling, you should bet only money you are able to lose. Always consult your budget!

ROULETTE

Roulette is the oldest casino game still in existence; the name is from the French *roué* meaning "wheel." The present form of the game comes from 1765 when a Parisian police official wanted a gambling game that would stop the cheating rampant in the city. Roulette first came to the United States through New Orleans in the early 1800s. Though many in Europe favor this game for its simplicity and elegance, it hasn't really caught on in the U.S.

The U.S. Roulette wheel consists of 38 numbered slots from 1 to 36, a zero, and a double zero:

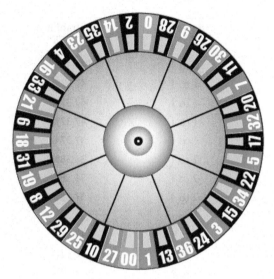

Figure 3C: American roulette wheel.

The table consists of every possible number as well as groupings called "outside" bets; these are number combinations.

		0		00
1-18	1st 12	1	2	3
		4	5	6
Odd		7	8	9
		10	11	12
Red	2nd 12	13	14	15
		16	17	18
		19	20	21
Black		22	23	24
Even	3rd 12	25	26	27
		28	29	30
		31	32	33
19-36		34	35	36
		2 to 1	2 to 1	2 to 1

Figure 3D: American roulette table.

Players make their bets, the dealer spins the wheel, then spins a ball in a track on the wheel. The ball eventually lands on a number and pay-offs are made accordingly.

The bets and pay-offs in Roulette consist of:

Bet	Pay-off	Probability
Red	1 to 1	47.37%
Black	1 to 1	47.47%
Odd	1 to 1	47.37%
Even	1 to 1	47.37%
1 through 18	1 to 1	47.37%
19 through 36	1 to 1	47.37%
1 through 12	2 to 1	31.58%
13 through 24	2 to 1	31.58%
25 through 36	2 to 1	31.58%
Sixline (6 numbers)	5 to 1	15.79%
First five (5 numbers)	6 to 1	13.16%
Corner (4 numbers)	8 to 1	10.53%
Street (3 numbers)	11 to 1	7.89%
Split (2 numbers)	17 to 1	5.26%
Single number	35 to 1	2.64%

Figure 3E: Roulette bets, payoffs and probabilities.

BLACKJACK

Gambling with cards spread steadily throughout Europe after Johann Gutenberg printed the first deck of cards in Germany in 1440. The game of "one and thirty" was first played in Spain sometime before 1570, whereas the game of 21 was first listed in America in 1875. Acceptance of the game in the U.S. was slow, so to stimulate interest, operators offered to pay 3 to 2 for any count of 21 in the first two cards. The pay-out was 10 to 1 if the 21 count consisted of the *ace of spades* and a *black jack*. Hence the name of 21 became Blackjack. The payout for any other win in Blackjack (or 21) is even money, that is, it pays 1 – 1.

A blackjack table consists of seven betting spots; however, it is not a group game. Each bettor is playing against the house; the number of players or where they sit has no outcome on the game. The decision to draw or not draw by any one player has no long-run effect on the other players. Many people mistake the goal of blackjack as trying to make a hand of 21 with designated card values. The actual goal is to be closer to 21, *without going over*, than the dealer. If you keep this goal in mind, it is a much more relaxing and enjoyable game.

The game begins with all players placing their bets. The dealer will deal the player and themselves two cards. The dealer will have one "up" card and one face down. Depending on the casino in which you are playing, the players' cards may be face up or face down. The cards with numbers are scored at face value, the Jack, Queen, and King are all valued at 10 points, with aces as the wild card; aces can have a value of 1 or 11 and you can go back and forth depending on the other cards in your hand.

Any hand of blackjack that involves an ace is called a soft hand. It is soft because you get to decide if you want to use it as a 1 or as an 11. A hard hand is then any hand of blackjack that does not have an ace. The count is a hard count and you do not get a choice. When you are asked if you would like to "hit," the dealer is asking if you would like for them to deal you another card. You respond "stand" if you do not want any more cards, and you "bust" if your count exceeds 21. A bust is an automatic loss regardless of what the dealer has or ends up with. A "push" occurs when the dealer and the player have the same total. The player's bet is returned so they do not win any money nor do they lose their original bet. You have the option to "double down" in which you double your original bet after looking at your first two cards. You are then dealt only one more. Good hands to double down on include hard 9 through hard 11. You may also choose to "split" your pair. If you are dealt a pair, you may place a second equal bet and play each as a separate hand. Though most casinos will not let you take a reference card with you, these cards are helpful in determining when you should choose each option.

DOUBLE DOWN DD   STAND S   HIT H   SPLIT P

SPLIT IF ALLOWED TO DOUBLE AFTERWARDS, OTHERWISE HIT H/P

SURRENDER IF ALLOWED, OTHERWISE HIT H/R

Your Hand:

Dealers Card Showing:										
	2	3	4	5	6	7	8	9	10	A
8	H	H	H	H	H	H	H	H	H	H
9	H	DD	DD	DD	DD	H	H	H	H	H
10	DD	DD	DD	DD	DD	DD	DD	DD	H	H
11	DD	DD	DD	DD	DD	DD	DD	DD	DD	H
12	H	H	S	S	S	H	H	H	H	H
13	S	S	S	S	S	H	H	H	H	H
14	S	S	S	S	S	H	H	H	H	H
15	S	S	S	S	S	H	H	H	H/R	H
16	S	S	S	S	S	H	H	H/R	H/R	H/R
17	S	S	S	S	S	S	S	S	S	S
A,2	H	H	H	DD	DD	H	H	H	H	H
A,3	H	H	H	DD	DD	H	H	H	H	H
A,4	H	H	DD	DD	DD	H	H	H	H	H
A,5	H	H	DD	DD	DD	H	H	H	H	H
A,6	H	DD	DD	DD	DD	H	H	H	H	H
A,7	S	DD	DD	DD	DD	S	S	H	H	H
A,8	S	S	S	S	S	S	S	S	S	S
A,9	S	S	S	S	S	S	S	S	S	S
2,2	H/P	H/P	P	P	P	P	H	H	H	H
3,3	H/P	H/P	P	P	P	P	H	H	H	H
4,4	H	H	H	H/P	H/P	H	H	H	H	H
5,5	DD	DD	DD	DD	DD	DD	DD	DD	H	H
6,6	H/P	P	P	P	P	H	H	H	H	H
7,7	P	P	P	P	P	P	H	H	H	H
8,8	P	P	P	P	P	P	P	P	P	P
9,9	P	P	P	P	P	S	P	P	S	S
10,10	S	S	S	S	S	S	S	S	S	S
A,A	P	P	P	P	P	P	P	P	P	P

Figure 3F: Blackjack reference card.

With the popularity of many movies involving gambling, especially the game of blackjack, many people set out to learn how to count cards in order to make millions. Keep in mind that these are movies and not likely scenarios. Look around the casino. Do you know how they can afford such extravagance? The house always wins. This is one of the main reasons that you should bet only money that you can afford to lose.

RED DOG POKER

The game of Red Dog was created in the 1800s in America. The game was originally known as In-Between, however, the original rules made it easy to cheat. When gambling was legalized in Las Vegas in 1931, the game of In-Between was reintroduced as Red Dog.

Of all the poker games possible in a casino, this is one of the easiest. Red Dog is a game of you versus the cards and not another player or even the dealer. A wager is placed then two cards are dealt. You may then place a second wager or stay with your original bet. You are never allowed to remove money from the table during a play of the cards. Your favorable outcome is that a third card has a value strictly between the two original cards. Pay-outs differ based on the spread of the count between the two original cards.

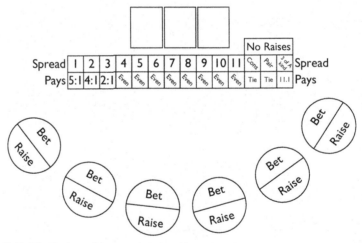

Figure 3G: Red Dog poker table.

The pay-outs for Red Dog Poker are shown in the table.

Spread	Pay-out
1 card	5 to 1
2 cards	4 to 1
3 cards	2 to 1
4 to 11 cards	1 to 1 (even money)

Figure 3H: Red Dog pay-outs.

There isn't much of a strategy for Red Dog, as this is a game of chance and should be played for enjoyment. Your best time to raise is if the spread is 7 or more. As this is a game of luck, you cannot control the game in any way or hope to skill your way to a win. Never bet more than you can afford to lose.

IN- AND OUT-OF-CLASS PROJECTS

1. As individuals, write a 5–7-page research paper on the effects of gambling addiction.
2. As individuals or in pairs, as directed by your instructor, create your own carnival-type game of chance.
    a. The initial bet must be $5.00 and the expected value must be positive but less than $0.50. (This is a fundraiser-type game so the operator must make money.)
    b. You must include at least 4 but no more than 8 events. Choose three to seven events that you want to occur and look at in more depth. You can make your last event "otherwise" to cover the less interesting events occurrences.
    c. Determine the probability of each event either experimentally or theoretically. Show your work or provide your data.
    d. Play your game 200 times (not to include the experimental probability calculations). Keep track of each outcome in groups of 50. That is, in the first 50 games, record how many times event 1 occurred, record how many times event 2 occurred, and so on.
    e. Make a bar graph and table showing the comparison between expected results and actual results in groups of 50.
    f. Write a paragraph detailing whether you won or lost money in this game and if the results were close to what you expected or not.
3. Go to a local casino or racetrack and write a three page paper based on your experience. Be detailed with sights, sounds, and everything else you encountered on your visit. Past trips to Las Vegas, Atlantic City, or local venues may be used.

CHAPTER FOUR

STATISTICS

9.

Jack remembered doing some types of statistical analysis in his coursework at TTU, but he didn't pay enough attention to remember it in detail. Every once in a while he thinks back to that class and realizes he should have asked more questions.

4.1 Jack read that a Gallup poll of 1,051 American adults shows that 32% of Americans say they have been spending less in recent months and 27% say they are saving more now and intend to make this their new, normal pattern in the years ahead. Describe the population, sample, population parameters, and sample statistics.

4.2 While Jack was reading the paper, his daughters were off in the corner giggling with phones in hand. Jack decides that he wants to determine the average number of hours per week that high school freshmen spend on cell phones. How would you apply the five basics steps of a statistical study to this issue?

Moving on to the sports section of the paper, Jack looks over the college basketball scores. On a particular day there were eight games in Conference USA with scores 81-62, 70-63, 66-57, 60-46, 70-64, 71-69, 98-77, and 70-63.

4.3 Find the mean, median, mode, range, and standard deviation for all the winning scores.

4.4 Find the mean, median, mode, range, and standard deviation for all the losing scores.

4.5 Find the mean, median, mode, range, and standard deviation for all the winning margins. The winning margin is the difference between the winning score and the losing score.

4.6 Provide a five-number summary for the winning scores, for the losing scores, and for the winning margin. Include a box plot for each five-number summary.

As Sally and Nicole are getting ready to apply to college, Jack and Diane help the girls get ready for their SATs. While the girls are working away, Jack decides to look up some information on this standardized test. He learns that in 2013, Critical Reading had a mean of 496 with a standard deviation of 115, Mathematics had a mean of 514 with a standard deviation of 118, and the Writing portion had a mean of 488 with a standard deviation of 114.

4.7 Suppose the school Nicole wants to go to requires scores in the 90th percentile for both Critical Reading and Mathematics. What score would she need to get on each portion of the test?

4.8 What percentage of the students are expected to score above 550 on the Writing portion of the exam?

4.9 Sally scored a 603 on the Mathematics portion of the exam. What score would she need on the Critical Reading to be in the same percentile? What score would she need on the Writing?

4.10 What percent of students score less than 350 on the Critical Reading? What percent of students score less than 350 on the Mathematics? What percent of the students score less than 350 on the Writing?

10.

In Week 7 of the 2013 NFL season, the twenty-eight active teams scored totals of 34, 22, 23, 31, 41, 45, 17, 3, 27, 30, 23, 21, 6, 15, 30, 27, 24, 31, 17, 16, 17, 16, 19, 13, 31, 39, 7, and 23 points. Jack, being a Seattle Seahawks fan, was happy to see that his team won that week. He did get curious about the point values, however.

4.11 Using 5-point bins (15–19, 20–24, etc.), make a frequency table for the data of points scored. Be sure to include columns for relative frequency as well as cumulative frequency.

4.12 Using 10-point bins (0 9, 10–19, etc.), make a frequency table for the data of points scored. Be sure to include columns for relative frequency as well as cumulative frequency.

4.13 Using the 5-point bins from problem 4.11, create a histogram of the data. Do the same using the 10-point bins of 4.12. Be sure to label your histograms completely.

Looking over the weather predictions for the playoffs, Jack finds the temperatures for the four teams of the NFC West for the coming weekend.

	Friday (Hi/Lo)	Saturday (Hi/Lo)	Sunday (Hi/Lo)
Phoenix	74/46	73/46	74/42
St. Louis	25/19	45/22	49/31
San Francisco	68/45	66/44	64/44
Seattle	49/34	49/35	50/35

4.14 Make a three-bar bar chart for these cities' high temperatures for the weekend.

4.15 Make a three-bar bar chart for these cities' low temperatures for the weekend.

As Jack skims the headlines of the newspaper, he tries to decide which articles he should read more of and which articles he can skip. For problems 4.16–4.18, help Jack determine if the study in the article should be believed or not. You must give a good reason for your opinion.

4.16 A study by the liberal Center for American Progress is designed to assess a new Republican tax-cut plan.

4.17 A study financed by Pfizer is intended to determine if its new cholesterol drug is better than the one put out by Merck and Company.

4.18 The Democratic Party polled 2,000 of its members to determine if their candidate for a U.S. Senate seat is likely to win against the Republican candidate.

9.

WHAT ARE STATISTICS?

If you are a sports fan you have heard many statistics in your life. But what exactly are statistics? First, statistics is the science of collecting, organizing, and interpreting data. Statistics is a field of study separate from mathematics, although they are generally found in the same department. Statistics are also the data that describe or summarize something. A common statistic in baseball is the RBI, or runs batted in. In football we may talk about the quarterback rating, another statistic. A basketball player may focus on his free-throw percentage. Statistics show up in more places than just athletics, however. A friend might encourage another friend to quit smoking so they "don't become just another statistic." This is referring to the number of deaths of smokers each year, a statistic. Anything that we can keep track of with numbers can be a statistic. Perhaps even the number of words per paragraph in this text could be a statistic … but not a very fascinating one.

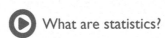 What are statistics?

COLLECTING DATA

Collecting data requires more than just writing some numbers down. Statistical studies are conducted in so many different ways that to list them here would be tedious. We will focus on the general picture involved in collecting data. We start with a couple basic terms. Our population may be all males aged 16–25, it may be all female heads of household, or it may be the television-viewing public. Whatever you are trying to study or make a claim about will be your population. The sample is a subset of the population from which you will get raw data. That is, you could not possibly speak to every male aged 16–25 in the United States in order to get their opinion, so you would use a sample of those individuals in order to make conclusions about the entire population. In order to carry out a statistical study, you must know what it is you want to measure. The population parameters are specific characteristics of the population that you are to estimate. Perhaps it is the number of professional athletes who do not have tattoos, or even the costs associated with attending college. Sample statistics are numbers of observations that summarize the raw data.

Regardless of the type of statistical study you are carrying out, there are five basic steps involved.

1. State the goal of your study precisely. A vague goal leads to vague data and many different interpretations of the results. We need to be very specific from the start.
2. Choose a representative sample from the population.
3. Collect raw data from the sample and summarize these data by finding sample statistics of interest.
4. Use the sample statistics to make inferences about the population.
5. Draw conclusions to determine what you learned and whether or not you achieved your goal.

Stating your goal is as you would expect; in science we say "make a hypothesis." For example, "How many sophomores on this campus have finished their math requirements or are currently finishing them?" This might let the administration know that they need to schedule more math classes for lowerclassmen, or possibly work with the advisors to encourage students to complete their math requirement earlier in their school career. For this particular study, we could get a list (with administrative permission) of all the sophomores registered at the school and ask them to complete an online survey. The students who answered would be the sample. Rarely would you have 100% participation, though this sampling method could end up being the entire population. Perhaps you could stand in the student union area for a few hours and ask passers-by first if they are a sophomore and then continue with your other questions. There are many ways of choosing the sample and collecting the data. Once you have your data, you will use some of the calculations that follow to make your inferences; this is where the math comes in.

SAMPLES

A representative sample is a sample in which the relevant characteristics of the sample members match those of the desired population. If you wanted to know about tattoos on professional athletes, you would probably not go to Comic-Con to do your polling. You will not find a true representative sample for your goal there. Once you find your appropriate representative sample, you will need to find a way to collect your data. There are many different options for doing this. Simple random sampling is when we choose a sample of items in such a way that every sample of a given size has an equal chance of being selected. You might let a computer program select your sample or even draw items out of a jar in order to have a simple random sample. Systematic sampling occurs when you select every 5th or every 50th member of the population to be a part of the sample. Convenience sampling happens when you take a poll in a classroom, or stand in a busy area of campus and ask those who pass by. This population is convenient, as they are close to you and easy to poll.

 Don't be fooled by bad statistics.

MEAN, MEDIAN, AND MODE

When most people talk about an average, they are speaking of the mean. To find a mean, we sum all the data values and divide by the total number of vales. For example, Peter had exam scores of 75, 87, and 88. To find his exam average we add these values 75 + 87 + 88 = 250 and then divide the result by 3, as there were three exams: 250/3 = 83.3. Peter's exam average is an 83.3. We find averages for many things not related to school as well. In Jim Zorn's first year as a Seattle Seahawk, he attempted 439 passes for a total of 2,571 yards. Though we do not have the number of yards of each completion, we do have the total yards and number of attempts. This takes us to the last step of the process to find that Zorn had an average of 5.9 yards per attempted pass in his first year of the NFL.

This is not the only way we can measure central tendency. For example, when researching a new neighborhood to move to, you will frequently find information about the median income. To find a median, you list all your data values in order from smallest to largest and then choose the middle value. If there is an even number of data values, average the two middle values. On a certain block in a neighborhood, the five household incomes were $23,145, $37,880, $39,475, $42,723, and $89,731. The mean household income is $46,590.80, a value higher than four of the five incomes listed. This is not very representative of the neighborhood, as one income was so much higher than the others. In determining the median income, we find it to be $39,475, the middle value. This is a more reasonable representation of the neighborhood. In a class of ten students, the scores on a quiz were 0, 0, 0, 0, 5, 7, 7, 8, 10, and 10 out of ten points. Three of the students were absent that day and another student did not answer any questions correctly. The mean, or average, is a score of 4.7 on the quiz. However, the missing three students have skewed the data. The median quiz grade would be a 6, the average of the two middle values 5 and 7.

Another measure of data is the mode. The mode is the value that occurs most often. If there are two values that occur the same number of times, greater than any other values, we say the data are bimodal. In our last example of quiz scores, both 7 and 10 occur twice. If we ignore the missing quiz scores in our last example, both 7 and 10 occur twice, the greatest number of times of anyone taking the quiz. We would say that the mode is 7 and 10 for this data set. On an exam given to a class of 16 students, the scores were 78, 79, 80, 80, 80, 80, 83, 84, 86, 89, 90, 92, 98, 98, 98, and 98. The mean (average) score is an 82, the median score (middle score) is an 85, and the mode is 80 and 98. That is, on average the students scored an 82, but if you were in the middle of the class, you probably scored an 85. More students scored an 80 and a 98 than any other scores on this exam.

STANDARD DEVIATION AND RANGE

Once we know the mean, median, and mode, we have some good information about the data. However, this is not all we need to know. Suppose the starting running backs in the NFC and AFC conference playoff games had 7 rushes, 9 rushes, 23 rushes, and 31 rushes for their teams in a single game. The mean number of rushes would be 17.5 per game, the median would be 16 rushes per game, and there would be no mode. In giving an average of 17.5 rushes per game, it makes all the running backs sound pretty busy even though two of those men touched the ball fewer than 10 times. We can describe the range of the data as the difference between the largest and smallest values in the data set. In this case, the range of rushing attempts is $31 - 7 = 24$. The range will give us the distance between the smallest and largest values and the median will give a middle value, but we need a way to tell how spread out (dispersed) the data really are. Standard deviation is that measure of dispersion. If the standard deviation is small, all the data are really close together. If the standard deviation is large, the data are very spread out. Finding the standard deviation is a little more labor intensive than finding the other values mentioned thus far, but it is also a more telling number. The standard deviation is the square root of the sums of the squares of the differences of the data values from the mean. Yes, that sounds confusing. In order to write a formula for standard deviation, we will need to use several letters of the Greek alphabet. Standard deviation is given by σ, where

$$\sigma = \sqrt{\frac{\Sigma(x_i - \mu)^2}{n}}.$$

We know that the radical symbol is for the square root. The capital letter sigma, Σ, is notation telling you to sum, or add, a set of values. Each value x_i is a member of the data set, the Greek letter mu, μ, is the standard representation of the mean, and the value n is the number of data values in your set. Let's go back to the running backs example. We know that the mean is 17.5 yards. We now subtract the mean from each of the data values and square the result.

Rushing Attempts	Difference from the Mean	Difference Squared
7	$7 - 17.5 = -10.5$	$(7 - 17.5)^2 = (-10.5)^2 = 110.25$
9	$9 - 17.5 = -8.5$	$(9 - 17.5)^2 = (-8.5)^2 = 72.25$
23	$23 - 17.5 = 5.5$	$(23 - 17.5)^2 = 5.5^2 = 30.25$
31	$31 - 17.5 = 13.5$	$(31 - 17.5)^2 = 13.5^2 = 182.25$

Figure 4A: Standard deviation calculation for rushing attempts.

Once we add the values in the last column, we will have the numerator under the radical of the formula for standard deviation; that sum is 395. We now divide 395 by 4, the number of running backs in our data set, to get 98.75. The standard deviation is the square root of this result: $\sqrt{98.75} = 9.94$. This rather larger standard deviation tells us that not only is the range large, but also the data values in that range are spread out.

Returning to the example of the class of sixteen students and their test scores, let's find the standard deviation.

Exam Score	Difference to μ = 82	Difference Squared
78	−4	16
79	−3	9
80	−2	4
80	−2	4
80	−2	4
80	−2	4
83	1	1
84	2	4
86	4	16
89	7	49
90	8	64
92	10	100
98	16	256
98	16	256
98	16	256
98	16	256
		Sum = 1,299

Figure 4B: Standard deviation calculation for student exam scores.

We now find $\sigma = \sqrt{\dfrac{1299}{16}} = \sqrt{81.1875} = 9$. Once again our standard deviation is 9, but what does it mean with this data set? If we look at scores that are 9 points below and 9 points above our mean, we find all the scores from 73 to 91. This leaves out only our top five exam scores and accounts for the other eleven scores. That is, 11 out of 16 scores were within one standard deviation of the mean. This tells us the data are grouped fairly closely, especially when the range is only 20. Notice that 11 out of 16 data points is the same as saying 68.75% of the data points were within one standard deviation.

QUARTILES AND THE FIVE-NUMBER SUMMARY

Knowing the mean, median, mode, standard deviation, and range are all helpful values. However, even this much information does not tell a complete story. In one example, running backs in the NFL, a standard deviation of 9 meant the data were spread out, whereas in another example, exam scores, a standard deviation of 9 meant the data were closely grouped. Dividing the data into quarters, or quartiles, will help us get a better view of the variation of the data. The lower quartile, or first quartile, divides the lowest one-fourth of the data from the upper three-fourths of the data. It is the median of the data values in the lower half of a data set. In our example of the sixteen exam scores, 80 is the first quartile. The middle quartile, or second quartile, is the overall median of the data. For our example the second quartile is 85. The upper quartile, or third quartile, divides the upper one-fourth of the data from the lower three-fourths of the data. It is the median of the data values in the upper half of a data set. In our example, 95 is the third quartile. We can now give a five-number summary of the data set of our class of sixteen exams:

Low	78
Lower quartile	80
Median	85
Upper quartile	95
High	98

Figure 4C: Five-number summary for student exam scores.

Every five-number summary will consist of the same set of numbers: low value, lower quartile, middle quartile, upper quartile, and high value. For many people, however, the numbers are just not enough.

BOX PLOTS

A box plot, or box-and-whiskers plot, is a visual representation of the five-number summary. Using a number line as reference, we mark the locations of each of the five values given in the summary. A line is drawn connecting the low value to the high value and a box is drawn from the lower quartile to the upper quartile as shown:

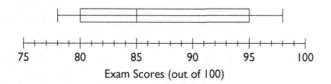

Figure 4D: Box plot of student exam scores.

In this box plot, it is much easier to see that the lower half of the data is grouped more tightly than the upper half of the data. As the median is not in the middle of the box plot, it is slightly to the left, we say that the data are skewed to the left. Recall that all the data values to the left of the mean were within one standard deviation. The box plot is another verification of the low variance of exam scores to the left of the mean. Also, there were five exam scores that were more than one standard deviation from the mean on the right of the data set. Again the box plot shows this variance.

NORMAL DISTRIBUTIONS

Normal distributions turn up in many of the things we can measure about humans, including many test scores. The normal distribution is a symmetric, bell-shaped distribution with a single peak. This peak corresponds to the mean, median, and mode of the distribution. In a normal distribution, about 68% of the data will be within one standard deviation, about 95% of the data will be within two standard deviations, and 99.7% of the data will be within three standard deviations. In our previous example, we saw that 68.75% of the data fell within the first standard deviation. We might say that the exam scores for these sixteen students follow a normal distribution.

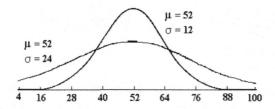

Figure 4E: Two normal distributions, both with mean 52. Notice the smaller standard deviation is taller and thinner whereas the larger standard deviation is shorter and wider.

To determine how many standard deviations any particular data value is away from the mean, we use the z-score, or standard score. A standard score is calculated by subtracting the mean from the data value and then dividing by the standard deviation. For our sixteen exam scores, a score of 86 is $\frac{86-82}{9} = .44$ standard deviations from the mean. Notice that data values larger than the mean will have positive z-scores and data values less than the mean will have negative z-scores.

We can use these z-scores to determine the percentile of the data point. The *nth* percentile of a data set is the smallest value in the set with the property that *n*% of the data values are *less than or equal to* it. A data value that lies between two percentiles is said to lie in the lower percentile. A standard score table is used to make the correlation between z-score and percentile and can be found easily online.

10.

TABLES, CHARTS, AND GRAPHS

The box plot is just one of many ways to represent statistical data visually. It is the most common way to represent the five-number summary, but how many people really know about quartiles (unless you're reading this book)? As you read through the newspaper, online articles, or even textbooks for your other classes, you are sure to find tables, charts, and graphs.

A frequency table is a method of showing the number of times any particular piece of data occurred, or its frequency. For example, in one class of 20 students, the final course grades were B C B B C C C B A F B F F A D B B C B A. These data are somewhat difficult to sift through, though not impossible. It would be quicker and easier to understand the grades if they were grouped according to letter grade and frequency. The grouping, in this case letter grade, will be called categories and make up the first column of our table. The second column will consist of the number of times each category occurred. For this example, our frequency table is

Letter Grade	Frequency
A	3
B	8
C	5
D	1
F	3

Figure 4F: Frequency table of final course grades.

In general a frequency table will have two columns. Additional columns may include the relative frequency of a category or cumulative frequency. The relative frequency is expressed as a fraction. The relative frequency of A's in this class is 3/20 = 0.15. All relative frequencies must add to 1, or 100% of the data. The cumulative frequency is found by adding the total frequency for a given category and all previous categories. The cumulative frequency for category

B is 3 + 8 = 11. That is, 11 of the students in this class received either an A or a B as their final grade. A larger frequency table is shown:

Letter Grade	Frequency	Relative Frequency	Cumulative Frequency
A	3	3/20 = 0.15 = 15%	3
B	8	8/20 = 0.40 = 40%	3 + 8 = 11
C	5	5/20 = 0.25 = 25%	3 + 8 + 5 = 16
D	1	1/20 = 0.05 = 5%	3 + 8 + 5 + 1 = 17
F	3	3/20 = 0.15 = 15%	3 + 8 + 5 + 1 + 3 = 20
Total	20	1 = 100%	20

Figure 4G: Frequency table of final course grades with added columns.

We could also show these data visually using either a bar graph or a pie chart. In a bar graph, the horizontal axis consists of the categories whereas the vertical axis is represented by the frequency, the relative frequency, or both (one on each side of the graph). Our bar graph will show frequencies only.

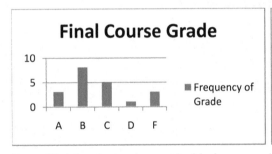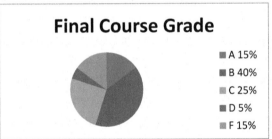

Figure 4H: A bar chart and pie graph of final course grades.

With this bar graph, it is easy to compare the number of A's with the number of F's and to see that there are more B's than any other grade in the class. Another popular method of representing data of this type is with a pie chart. Generally, we use relative frequencies in our pie charts in order to show percentages of the data.

Any time that you are making a graph or chart, it is important to keep some key information in mind. First, use an appropriate title for your graph. If you were reading the newspaper and saw a pie chart titled "Pie Chart 1," you wouldn't have any idea what the data were representing. Second, always label your axes and indicate units. If necessary, include a legend as shown in the pie chart. With the help of technology, the creation of these graphs is much easier than it used to be.

Not all of our data will be the same. Some data, such as class grades, are considered to be qualitative data, because they describe a quality or category. Other data, such as ages, are quantitative data because they represent a count or measurement; a quantity. For qualitative data we will use bar graphs and pie charts. For quantitative data we use primarily histograms and line charts. A histogram looks a lot like a bar graph but has one major difference: a bar graph has space between the categories because they are distinct, and a histogram has no spaces between the bars because the data run consecutive. Let's focus more on this idea with an example.

Recall the example of our class of sixteen with exam scores 78, 79, 80, 80, 80, 80, 83, 84, 86, 89, 90, 92, 98, 98, 98, and 98. In order to represent these data using a histogram, we first need to bin our data. To bin data we group it together in a meaningful way. For this example we will use bins of 75–79, 80–84, 85–89, 90–94, and 95–100. A bin is just a range of values, whether it be ages, heights, or in this case, exam scores. Next, we determine the frequency of scores for each bin and arrange it in a table as before.

We now use the histogram option of Microsoft Excel, or another spreadsheet program, to create our visual.

Exam Scores	Number of Students
75 – 79	2
80 – 84	6
85 – 89	2
90 – 94	2
95 – 100	4

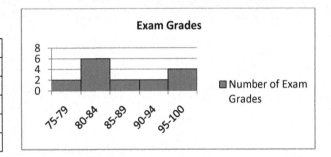

Figure 4I: Frequency chart of student exam scores using binning with corresponding histogram.

A histogram is a bar chart for quantitative data categories. The bars have a natural order and the bar widths have specific meaning. A line chart shows the data value for each category as a dot, and the dots are connected with lines. For each dot, the horizontal position is the center of the bin it represents and the vertical position is the data value for the bin. For the same sixteen exam grades, we have the following line chart.

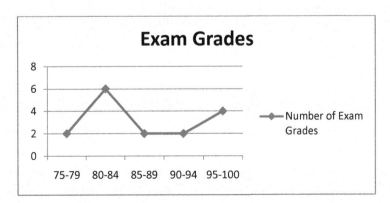

Figure 4J: Line chart of student exam scores.

In a histogram, you can make comparisons much easier. In a line graph you can look more at trends. Both graphs represent the exact same information; it all depends on what you are trying to show.

These are just a few of the many ways that we can represent data visually. We can also use multi-bar graphs, stacked graphs, three-dimensional graphs, pictographs, and geographical graphs. Examples of these are shown. One thing to keep in mind for all of our graphs and charts is the scaling. It is easy to misrepresent your data by changing the scale on either axis.

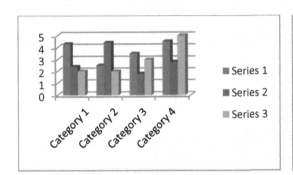

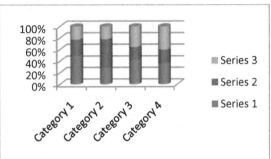

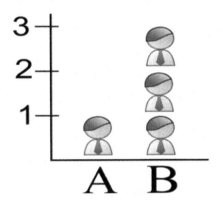

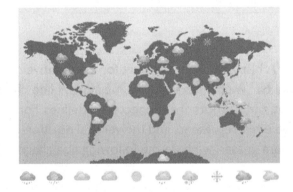

Figure 4K: Generalized multi-bar chart, stacked chart, pictograph, and geographical graph.

SHOULD YOU BELIEVE A STATISTICAL STUDY?

Many people see numbers and automatically assume that they must be correct. That is, the numbers support whatever the researchers suggest they support. However, in order to be a good consumer and a responsible adult, we should always make sure that the study is believable. You do not have to be an expert in the field in order to determine if you should believe a study; all you need to do is take a moment and consider these eight guidelines.

First, identify the goal, population, and type of study. If these aspects of the study are not easily recognizable, it will be difficult to consider other parts of the study. Next, we must always consider the source. In a poll taken by FOX news, it is most usually leaning more toward the right politically. Did they poll only their viewers? Then the data may be skewed more toward what they wanted to show anyway. This happens with many different organizations and program viewers on a daily basis. We will consider any potential bias in the study next. Selection bias occurs whenever researchers select their sample in a way that tends to make it unrepresentative of the population. For example, maybe you want to find the average height of students on your campus. Asking only the athletes for their heights will not truly represent the heights of all the students on campus. Participation bias can also occur. This is primarily with surveys and polls. Generally, only those with a strong opinion will answer a voluntary poll. These people with strong opinions may not represent the entire population.

Our fourth consideration is to look for problems in defining or measuring the variables of interest. Suppose a poll in early February is asking men how much they love their significant other. Will this be measured statistically by the amount of money spent on Valentine's Day? Will it be measured on a 10-scale? Love is difficult to measure, as it means something different to each person. Next, we need to look for other variables that may confuse (confound) the study. If you are studying the effects of exercise on heart health, you may find that one population has perfectly fine heart health even though they rarely work out, whereas the other population works out constantly and they have the same level of heart health. A confounding variable may be the diet of the individuals in the study. If the first population eats a heart-healthy diet but rarely has time to exercise, their hearts may be just as healthy as individuals who work out for two hours per day but eat all their meals at fast food restaurants.

In any statistical study you should consider the setting and wording in the survey. Ask a group of students leaving the financial aid office on the first day of school if they found the process to be quick and easy and you may get a radically different answer than you would if you asked in the middle of the semester instead. Asking a group of people if they cheat on their significant other as part of a Valentine's Day poll may not be accurate for a variety of reasons; some being that they are standing next to their partner and cannot give a truthful answer, or they know the individuals doing the polling and they are afraid to give a truthful response, as it could get back to their partner. You should always check that the results are presented fairly. Does the graph or chart accurately reflect the study or did they tweak the scaling to make it look like the data tell you something different? Did they provide a confidence interval or margin of error? All of these things need to be considered when evaluating the results. Finally, use your common sense. Did the study achieve its goal? Do the conclusions make sense? Do the results have any application to anything useful in life?

IN· AND OUT·OF·CLASS ASSIGNMENTS

1. In small groups, determine two or three items you would like to measure for the members of your class. (Hint: weight is not a good idea if you have very many females in the class.) Consolidate these items into one survey for each member of the class to fill out. Once the surveys are completed, find the mean, median, and mode for each item your group suggested. Next, find the standard deviation. If appropriate, also separate your data into categories such as male and female and recalculate the mean, median, mode, and standard deviation. Provide appropriate tables, charts, and graphs as well.

2. In small groups, determine a small statistical study that you can carry out on your campus. Make it a significant study in relation to helping the students of your school (data on heights are not helpful unless you find the doorways to be too low). Get all necessary approval from the administration of your school to conduct a statistical study, and then follow the process of statistical studies. Write a 5-page paper that summarizes your study goals, process, calculations, and conclusions. Support your information with appropriate tables, charts, and graphs.

3. Look in a newspaper, magazine, or journal to find a poll. Analyze the poll in detail.

CHAPTER FIVE

MODULAR ARITHMETIC

11.

As a child, Jack was always fascinated by numbers. After he learned to count, he would count things at every opportunity. This familiarity with numbers helped him pick up arithmetic quickly. The one thing that really stumped him was clock arithmetic. He just couldn't figure out why five hours after 8 o'clock wasn't 13 o'clock. His dad sat him down and explained the basics of clock arithmetic and a whole new world opened to Jack. He was on a quest to find other instances in life where we use this type of arithmetic. To remind him of his search, Jack insisted on wearing only analog watches and never digital.

5.1 Jack has a school project due and his math teacher tells him it is due in exactly 216 hours. If it is currently Monday at 10:30 AM, when is Jack's project due? Show your work.

5.2 At the beginning of Jack's first final exam as a college student, his professor says, "It has been a fun 2,616 hours this semester but it all ends here." Find the number of days in the semester. Show your work.

5.3 Jack is counting down the days until he takes a road trip with his friends. It is currently Wednesday and they are leaving in 45 days. On what day of the week will they leave? Show your work.

5.4 Jack loves to watch football. He saw an interview on December 29, 2013 with an NFL star who said, "It has been a long road these last 158 days since training camp started … ." On what date did training camp start for that team that year? Show your work.

5.5 Jack's birthday is April 2, 1972. His birthday in 2013 was on a Tuesday. On what day of the week was Jack born? Show your work.

5.6 What is the next year that Jack's birthday will be on a Tuesday? Show your work.

12.

W hile in college, Diane worked in the library. Jack would go visit her and help whenever he could just to spend time with her. One day the library scanner broke and so all checked books had to be input to the system by hand.

5.7 A student wanted to check out a book but the ISBN was missing the last digit. All Jack could make out to read to Diane was 0-385-72222-□. Find the missing check digit. Next, convert this to the new 13-digit ISBN by finding the new check digit. Find the title, author, and publisher of this book.

5.8 Jack decided to read some urban legends but this ISBN was smudged. He could make out the digits 0-7?07-6211-2. Find the missing digit. Convert this ISBN to the 13-digit format and find the new check digit.

One afternoon as Jack was putting groceries away, he noticed that two UPCs were similar. He decided to analyze a few products from his pantry.

5.9 The UPC on one of Jack's breakfast items was 0 30000 01040 □, where the check digit was ripped off when he opened the package. Find the missing check digit. Show your work.

5.10 Another one of Jack's breakfast items had the UPC ripped in two when he opened the package. Putting it together like a puzzle he discovered the numbers 0 30000 06□20 8. One digit was missing—find it. Who is the manufacturer of Jack's breakfast items? Show your work.

5.11 Surprised to find a UPC on a new pair of shoes, Jack read the number quickly as 0 85681 78969 3. He threw the box away before he realized that it couldn't possibly start with a 0, as shoes are not food. Find the correct starting digit of the UPC. Show your work.

Christmas is coming soon and Jack really hates crowds. He decides to do as much shopping online as he can.

5.12 Before Jack buys gifts for his loved ones, he decides to pay a couple of bills. While paying his utilities on the utilities website, he is asked to enter his bank routing number. He enters the digits 100700327, but the computer tells him it is incorrect. He knows he made a transposition error somewhere; find the two digits he switched. Show your work.

5.13 Now that his bills are paid, Jack begins searching for the "perfect" gifts for his beautiful daughters. He is ready to check out but is too lazy to go get his credit card because he is pretty sure he remembers the number. He enters 4323 1234 5678 912□ and then blanks on the last number. If he has remembered the first 15 digits correctly, what is the last digit? Show your work.

5.14 In order to make a few extra bucks before the holiday season, Jack is selling some old action figures on a local internet site. The buyer shows up at an agreed-upon meeting location with money order in hand excited about owning his new (old) Transformers still in the box. Jack looks at the money order and knows it is a fake. The number listed was 73521943815. How did Jack know it was not an authentic money order? What should the last digit have been? Show your work.

13.

A t Jack's architecture firm all employees must have identification badges. This badge has the form LNNN-NNNN-NNNN where L represents a letter and N represents a numeric digit. The first section, LNNN, identifies the person's last name using the Soundex Coding System. The first three numbers of the middle section, NNN, identify the employee number. The last digit of the middle section along with the last section, N-NNNN, identifies the person's sex and date of birth using the Florida driver license scheme.

5.15 Find the employee badge identification number for Jill Widman, born July 7, 1970 who was the 34th person hired by the firm.

5.16 Find the employee badge identification number for Roberto Russell, born February 16, 1968 who was the 145th person to join the firm.

5.17 If you were hired by Jack's architecture firm as the 369th employee, what would be your badge identification number?

When Jack and Diane bought the new car, they needed to report the VIN to the insurance company. Jack decided to take a closer look at all the VINs for the automobiles they had.

5.18 Jack decides to use his knowledge of check digits to find the details of his car. The VIN is 1C3EL45R12N305884. Find the make, model, year, and all other information for this particular vehicle.

5.19 While writing down the values Diane is reading to him, Jack couldn't make out if she said 2CNDL63F576091290 OR 2CNDL63F576091209 because nine and ninety sound similar. Assuming all other digits are correct, determine whether Diane said 90 or 09 as the last two digits.

5.20 Jack knows this system works on cars, trucks, and SUVs, but he wonders if it works for motorcycles; after all, they are vehicles too! He writes down the VIN from his motorcycle, leaving the 9th digit blank, to see if the same calculations were made to find it. The VIN he wrote was WB101790?9ZW16784. Find the missing check digit and find the details of this motorcycle.

11.

MODULAR BASICS

Modular arithmetic was first studied in modern times by Carl Friedrich Gauss and was published in his *Disquisitiones Arithmeticae* in 1801. Modular arithmetic is a system of arithmetic in which numbers do not continue forever, but instead start over once we reach a specific number. For this reason, we can also think of modular arithmetic as the arithmetic of remainders. In modulo 10, our numbers would start over at 10. In modulo 14, the numbers would start over at 14. Most of the time, modular arithmetic uses the digits 0, 1, 2, 3, ... , n-1 if we are looking at a modulo *n*. That is, we start at zero and count up to one before the modular number. In modulo 7 we have digits 0, 1, 2, 3, 4, 5, 6 and then once we reach 7 it returns to 0 as division of 7 by 7 yields a remainder of 0. Counting to 10 in modulo 7 would give us: 0, 1, 2, 3, 4, 5, 6, 0, 1, 2, 3. There is one big exception to the starting at zero rule and that is with a clock. A clock uses modulo 12 with values 1, 2, 3, 4, 5, 6, 7, 8, 9, 10, 11, and 12.

CLOCKS AND CALENDARS

Other than clocks, why would we choose to use a modular arithmetic? The principles of modular arithmetic arise in any part of our world that is cyclic in nature. Consider the calendar; our seven days each week start over for the next week. Our 365 days start over each year. The seasons are cyclic, tides are cyclic, the cycle of the moon is as well. We see modular arithmetic every day and yet very few people are familiar with it. Suppose it is currently 11:30 AM. We know in twenty-four hours it will once again be 11:30 AM. We can think of our days as arithmetic modulo 24. However, if it is still 11:30 AM, we know that in twelve hours it will also be 11:30, this time PM. We can think of clock arithmetic in modulo 12 or modulo 24 depending on what it is we are actually trying to determine.

Suppose it is 1:00 PM and you have seventeen hours to turn in your term paper. We know that in twelve hours it will once again be 1:00, this time in the morning. Add another five hours (12 + 5 = 17)

and our paper will be due at 6:00 AM. This time, suppose we are studying for an exam that occurs in 77 hours. It is currently 11:00 AM on Saturday. What time will the exam take place? We know that we can break the 77 hours into three groups of 24 with five left over. That is, $77 = 24 \cdot 3 + 5$. This tells us the exam is three days and five hours from the current time. Our exam is Tuesday at 4:00 PM.

Figure 5A: A clock with incorrect Roman numeral representation of the number 4 and another with the correct representation.

We can also count backwards using the arithmetic of remainders but we need to be careful in doing so. Pay attention to the details. If it is currently 5:00 PM, nine hours ago it was 8:00 AM. Suppose it is 3:00 PM. What time was it 20 hours before? We know that it was 3:00 AM twelve hours ago. Subtract another eight hours ($20 = 12 + 8$) to find it was 7:00 PM. If it is currently 12:00 noon on Sunday, 114 hours ago it was four days earlier, AM rather than PM, and then subtract six more hours. Mathematically this can be represented as $114 = 24 \cdot 4 + 12 \cdot 1 + 6$. Therefore it must have been Tuesday evening at 6:00 PM. (The four days take us to Wednesday, but then the twelve hours take us to 12:00 midnight and six hours previous rolls us back into Tuesday.)

We can do similar calculations with the days of the week. Suppose it is currently Saturday. In twelve days it will be Thursday. Mathematically we think of this as $12 = 7 + 5$. The seven moves us forward one week, landing back at Saturday. We then count forward five days to arrive at Thursday. Suppose it is Monday and you are sitting in class reading the syllabus. As part of your grade you are assigned to complete community service for 45 straight days at a facility of your choosing. You decide to begin your service on a Thursday; on what day of the week will your community service end? We once again turn to our arithmetic of remainders to find that $45 = 6 \cdot 7 + 3$. That is, your community service will last for six weeks and three days. Your service will end on a Sunday. Your best friend is taking the same class and ended his service on a Wednesday. On what day of the week did he begin his community service? It was six weeks and three days previous, so he began on a Sunday.

Calculating years is nearly the same as weeks and days; we just have to make corrections for leap years. It takes 365.2524 days for the earth to rotate around the sun. We generally make the estimate as 365.25 years but the entire decimal is important. Many people are familiar with the idea of leap year, a year in which an extra day (February 29) is added to the calendar to keep the balance. The extra day occurs every four years, offsetting the quarter of a day in 365.25 days. How do we handle the 0.0024? Leap year occurs every four years except in the instance of a century marker. Leap year occurs only in century years that are divisible by 400. That is, the year 2000 was a leap year but the year 1900 was not.

January 1, 2014 fell on a Wednesday. We can determine that January 1, 2013 was 365 days previous, where $365 = 52 \cdot 7 + 1$, so it was 52 weeks and one day earlier on a Tuesday. This calculation didn't involve a leap year. How can we determine on which day of the week January 1, 2010 fell on? (Using Google is cheating.) We need to determine the total number of days to count back. We know 2013 had 365 days, 2012 had 366 days, 2011 had 365 days, and 2010 had 365 days as well. We now need to count back 1,461 days. Doing the calculations we find $1461 = 208 \cdot 7 + 5$. So we go back 208 full weeks and five days. Five days before Wednesday is a Friday. Notice that even though it was four years earlier, we had to count back five days due to leap year.

CHECK DIGITS

Check digits are a way of making sure the identification numbers we see and use every day are input correctly. Check digits allow us to detect many different types of errors when entering a UPC code by hand at the grocery store, an ISBN by hand at your local bookstore, or even when entering a tracking code for your latest online purchase through a shipping website. To find check digits we use modular arithmetic. It seems as though we use numbers for everything in our society. We have bank routing numbers and account numbers, we have credit card and debit card numbers, every package shipped has a tracking number, money orders, books, foods, medicines, the list goes on and on. With so many systems in place, it should not be surprising to find that there are many different methods of determining the check digit as well. Some identification numbers use simple division, while others use a method of weighting. Details for each algorithm can be found by type of identification number in the following passages.

12.

ISBN

Nearly every book you have ever picked up has an ISBN. ISBN is an abbreviation for International Standard Book Number; saying the book has an ISBN number would therefore be a redundancy. The ISBN was created to help catalog and identify published works in a unified approach. Many countries had their own method of identifying publishers and titles, but this is the first system to be used world-wide. Originally developed in the mid-1960s, it was not in full use until the late 1970s. For many decades ISBNs were ten digits in length. In 2007, a new system was integrated that uses thirteen digits. On many books published around that time, you can find both the 10-digit and 13-digit ISBNs.

(5B)

Figure 5B: A ISBN 10-digit and 13-digit.

The numbers for an ISBN are not randomly selected. They are assigned by a registering agency for the country of publication. Each ISBN contains information about the group (language or country of origin), the publisher, the title, and the last digit is a check digit. First, let's discuss the 10-digit ISBN and the algorithm for finding the check digit. *The Complete Works of Edgar Allan Poe* has ISBN 0-394-71678-7. The zero identifies this as published in the United States in English, as would a one in the starting position. The 394 identifies the publisher as Alfred A. Knopf, now part of Random House's Knopf Doubleday group. The 71678 is the specific number given to the particular book title and the 7 is the check digit. As all else is assigned, how do we find the check digit 7?

The ISBN 10-digit uses a weighting system of 10, 9, 8, 7, 6, 5, 4, 3, 2, and 1. That is, multiply the first digit by 10, the second digit by 9, and so on. The sum of these should be congruent to 0 mod 11. That is, the sum should be evenly divisible by 11. Let's check our example:

$$10(0) + 9(3) + 8(9) + 7(4) + 6(7) + 5(1) + 4(6) + 3(7) + 2(8) + 1(7)$$
$$= 0 + 27 + 72 + 28 + 42 + 5 + 24 + 21 + 16 + 7 = 242$$

We know that $242 = 11 \cdot 22 + 0$, with the remainder of zero, we know this is a valid ISBN. Suppose we have only the first nine digits and we wanted to find the check digit. In this sum we would have $235 + C$, where C is the check digit. We need to determine what C should be in order to make the sum evenly divisible by 11. In this case we already know that $C = 7$.

Let's consider another book, *Classics of Mathematics*, with first nine digits of the 10-digit ISBN given by 0-02-318342-?. We need to find the check digit:

$$10(0) + 9(0) + 8(2) + 7(3) + 6(1) + 5(8) + 4(3) + 3(4) + 2(2) + ?$$
$$= 0 + 0 + 16 + 21 + 6 + 40 + 12 + 12 + 4 + ? = 111 + ?$$

We need to find a value for ? so that the sum is divisible by 11. The next multiple of 11 after 111 is 121. We need a check digit of 10 to make this a correct ISBN. The word "digit" is very specific, however. A digit has a single value, 0–9, so 10 is not a digit, it is a two-digit number. As we need a check digit of 10, but cannot use a two-digit number, we use an X in its place. Our ISBN therefore is 0-02-318342-X. The check digit of 10 is the only one to receive a special symbol; all other check digits in mod 11 will be single-digit values and therefore need no replacement.

The ISBN 13-digit is actually an easy conversion from the 10-digit original. For the most part we place "978" in front of the original 10-digit ISBN and then calculate a new check digit. The middle values of both the 10-digit and 13-digit ISBNs should be identical in identifying group, publisher, and title of the work. Let's now convert *The Complete Works of Edgar Allan Poe* to the 13-digit ISBN. The new ISBN would be 978-0-394-71678-?, as we still need to find the check digit. Rather than use the weighted system of the 10-digit ISBN codes, the newer 13-digit ISBN has a new weighted system. We multiply alternately by 1 and 3 and the resulting sum must be divisible by 10. That is, the 13-digit ISBN is congruent to 0 mod 10. Let's find our missing check digit:

$$1(9) + 3(7) + 1(8) + 3(0) + 1(3) + 3(9) + 1(4) + 3(7) + 1(1) + 3(6) + 1(7) + 3(8) + 1(?)$$
$$= 9 + 21 + 8 + 0 + 3 + 27 + 4 + 21 + 1 + 18 + 7 + 24 + ? = 143$$

We need to find a value for ? that when added to 143 makes a value evenly divisible by 10. The obvious choice is $? = 7$. Having a check digit that makes a sum divisible by 10 is much easier than finding a check digit to make a sum divisible by 11, as we all know any number evenly divisible by 10 must end in a zero. We can then focus our attention on only the ones place of our sum in order to make the check digit work out. We can do similar calculations for *The Classics of Mathematics*, but let us do it in a slightly different way. We know that the digits in the odd place values are multiplied by one and the digits in the even place values are multiplied by three. For the 10-digit ISBN 0-02-318342-X, we tag the 978 in front and remove the check digit from the end replaced by a question mark:

$$1(9 + 8 + 0 + 3 + 8 + 4 + ?) + 3(7 + 0 + 2 + 1 + 3 + 2)$$
$$= 32 + ? + 3(15) = 77 + ?$$

In order to make this sum congruent to 0 mod 10, we know that ? = 3. Our new 13-digit ISBN is 978-0-02-318342-3.

UPC

When grocery stores switched to UPC codes and electronic code readers in the 1970s, it saved the grocery industry over $17 billion in the first year alone. The use of UPCs saves the grocery industry over $30 billion per year now. How could one simple idea save so much money? Before the 1970s, and in rural America even now, clerks at stores had to enter the price of each item by hand. Every item first had to be priced before being shelved, then the price had to be read and entered into a basic cash register (glorified adding machine that could store money). The stock person could have tagged the wrong price on the item, the clerk may have read the price incorrectly, the clerk may have entered the numbers incorrectly, or the price tag itself may have fallen off the product. With a uniform code system, the product can be entered into the store's system with accurate pricing. You would no longer need someone to tag every single item and you no longer need to rely on the clerk entering the price correctly. Both of these save time and money for the store owner.

Figure 5C: UPC label.

A UPC consists of a bar code and associated numbers. The numbers uniquely describe the bar code, and the bar code uniquely matches the numbers. The first digit generally refers to the type of item such as food, medicine, coupons, and so on. The next group of digits identifies the manufacturer of the product, the third group of digits specifies the product, and the last digit is a check digit. The UPC algorithm for finding the check digit is very similar to the method of the 13-digit ISBN with one small change. For the UPC check digit, multiply the digits in the odd place value by 3 and the digits in the even place values by one. The sum of these products should be divisible by 10; that is, the UPC algorithm has a check digit that is congruent to 0 mod 10. This method of determining the check digit detects all single-digit errors and will catch up to 89% of all transposition errors (entering 12 as 21).

The UPC for Coke Zero is 0 49000 04256 6. To use the algorithm in checking to make sure the 6 is a correct check digit, we calculate the following:

$$3(0 + 9 + 0 + 0 + 2 + 6) + 1(4 + 0 + 0 + 4 + 5 + ?)$$
$$= 3(17) + 13 + ? = 64 + ?$$

It is straightforward to see that indeed the 6 in place of the ? makes the sum evenly divisible by 10.

Some products, such as soda pop, have a small surface area that makes it difficult to place the full 12-digit UPC on the product itself. In this case, the UPC can be converted to a short code that essentially eliminates the zeros at the end of the manufacturer code and the zeros that begin the product code. A can of Dr Pepper has UPC 0 783150 4, whereas the 12-pack box of Dr Pepper has UPC 0 78000 08216 6. The UPCs are not identical, as they are not the same product; a 12-pack has a different pricing than a single can or bottle, but you can see the similarity of the manufacturer code in both UPCs. We will use only the 12-digit UPC for the purposes of this text.

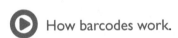 How barcodes work.

IDENTIFICATION NUMBERS

The most recent 13-digit ISBN and the standard UPC use the same weighting system to find check digits for their coding. However, this is not the only system used. The U.S. banking system, along with many others, want an even more reliable way of encoding numbers so that money gets to where it is supposed to go. Each bank has a 9-digit identification number, also called the routing number. The first eight digits refer to a specific bank and the last digit is a check digit. In order to find the check digit, we multiply each digit by 7, 3, 9, 7, 3, 9, 7, 3, and 9; the sum must be divisible by 10. For example, Pioneer Bank has routing number 312270463. Let's test the weighted system:

$$7(3) + 3(1) + 9(2) + 7(2) + 3(7) + 9(0) + 7(4) + 3(6) + 9(?)$$
$$= 21 + 3 + 18 + 14 + 21 + 0 + 28 + 18 + 9? = 123 + 9?$$

We need to find a value for ? so that when multiplied by 9 the ones digit is a 7, this is because 123 added to something ending in a 7 will be evenly divisible by 10. Consider all the single-digit multiplies of 9: 9, 18, 27, 36, 45, 54, 63, 72, and 81. Notice that each multiple has a unique digit in the ones place. This tells us that we must add 27 to 123 for a sum of 150. In order to make 27 with a factor of 9, our missing digit must be a 3. Notice that is exactly what was in the routing number to begin with.

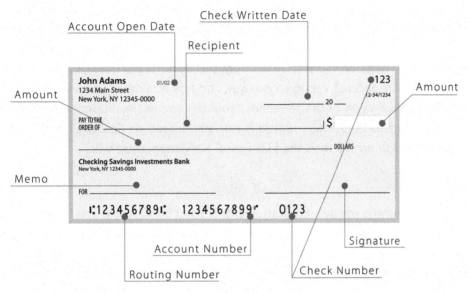

Figure 5D: Sample check.

Some banks have so many branches that they have different routing numbers in different states. In California, Chase Bank has routing number 32227162?. Find the missing check digit.

$$7(3) + 3(2) + 9(2) + 7(2) + 3(7) + 9(1) + 7(6) + 3(2) + 9(?)$$
$$= 21 + 6 + 18 + 14 + 21 + 9 + 42 + 6 + 9? = 137 + 9?$$

It must be that ? = 7 because 137 + 63 is congruent to 0 mod 10. Using this system, online bill-pay systems can make sure that the account you are specifying is for an actual bank.

U.S. Postal Service money orders have an 11-digit identification number in which the first 10 digits tell about the money order itself and the last digit is a check digit. Money orders use a very rudimentary system for finding the check digit that does not catch transposition errors, but will catch most single-digit errors. The algorithm to find the check digit is to add the first ten digits and divide by nine. The check digit is the value of the remainder of your division. For example, a money order has identification number 6312472907?, where we need to find the value of the check digit ?. We sum the digits to get 41. The remainder when 41 is divided by 9 is 5, 41 = 9 · 4 + 5. This means the money order would have a full identification number of 63124729075.

American Express and VISA traveler's checks use a similar system, though it is not exactly the same. The check digit is chosen so that the sum of the digits, including the check digit, is evenly divisible by 9. Rather than using the remainder as the check digit, the check digit is a value needed to make the sum evenly divisible.

Airline tickets, package delivery companies, and some rental car companies assign a check digit based on the division of the identification number itself by 7. If your rental car identification code is 4386219?, we must divide 4,386,219 by 7 and place the remainder in for the ?. In this case the remainder is 5. The full identification number for this rental car would be 43862195. To complete this task quickly, on a calculator rather than by hand, divide the number by 7. Next, subtract off the whole-number value to leave only the decimal portion, this is the remainder. This is the remainder in decimal version, however, and not a number you will choose. To find the actual whole-number value of the remainder, multiply the decimal portion by 7. In this way you find the check digit to be 5. This method is not guaranteed to detect all errors, but it will catch more transposition errors than the U.S. Postal Service method will.

Codabar

One of the most efficient error-detection methods is used by all major credit card companies along with many other institutions. Each credit card is assigned a 16-digit number; the first 15 digits are for the account and the last digit is a check digit. This is how a computer will know if you entered the wrong credit card number for an online purchase. The method is not as straightforward as multiply and add, but we definitely want a better error-detection method for our credit cards.

Figure 5E: Generic credit card.

A bank issues a credit card with identification number 3125 6001 9643 001?, where the ? is the check digit to be determined. First, we add the digits in the odd place values, and then double the results: $2(3 + 2 + 6 + 0 + 9 + 4 + 0 + 1) = 2(25) = 50$. Next, count the number of digits in the odd place values that are greater than 4 and add this to our total. Only the digits 6 and 9 exceed four, so we add 2 to 50 to get 52. Finally, we take our sum of 52 and add in the digits in the even place values: $52 + (1 + 5 + 0 + 1 + 6 + 3 + 0 + ?) = 52 + 16 + ? = 68 + ?$. In order for this to be a valid credit card number, the sum must be divisible by 10; therefore the missing digit must be a 2. When this credit card is issued, it will have number 3125 6001 9643 0012.

PERSONAL DATA

Given a social security number, say 043-25-1234, what does it tell us about that particular individual, if anything? From this number we know that the individual received their SSN from the state of Connecticut, but little else. Suppose your friend from Illinois shows you their driver license with number 1225-1637-2133. This time we can find the date of birth, sex, and quite a few things about the individual's last name. A social security number encodes no personal data about the individual. The number is generated based solely on where and when the number was requested. At the other extreme the Illinois driver license encodes a great deal about the individual. Agencies with large databases find it more convenient to store personal information in the identification number rather than just as a reference number like the Social Security Administration. Keeping in mind that the Social Security Administration was around before computers will allow us to forgive the rudimentary use of that number identifier. The National Archives, where census information is stored, genealogical research centers, the Library of Congress, and many state motor vehicle departments choose to encode our personal data into our identification number.

One way of coding personal information is the Soundex Coding System. We can use this to determine the first four characters of a driver license number. As our example, we will use my last name of Johnson.

1. Delete all occurrences of *h* and *w*. I am now Jonson.
2. Assign numbers to the remaining letters as follows:
    a. Vowels, including *y*, have value 0.
    b. *b, f, p, v* have value 1.
    c. *c, g, j, k, q, s, x,* and *z* have value 2.
    d. *d* and *t* have value 3.
    e. *l* has value 4.

    f.   *m* and *n* have value 5

    g.   *r* has value 6.

3. If two or more letters with the same numerical value are next to each other, omit all but the first. For example, Schneider becomes Sneder.

4. Delete the first character of the original name if it is still present. I am now onson.

5. Delete all occurrences of vowels, as they have value 0. I am now nsn.

6. Keep only the first three digits corresponding to the remaining letters; add trailing zeros if fewer than three letters remain. Precede the three digits with the first letter of the last name. Giving values to nsn and using the first letter J of my last name, my Soundex code would be J-525.

Notice that all spellings of Johnson, including Johnsen, Jonsen, and Jonson, would have the same Soundex code. This is extremely helpful for airline reservations in which the gate agent may not be familiar with one particular spelling of a last name. Perhaps a state patrol officer rushes to put in a last name spelled incorrectly due to the pouring rain. In either instance, the computer is searching for J-525 and not the actual spelling.

We have the name encoded for our driver license, so how can we throw in the date of birth and sex? There are many systems for encoding this data but we will focus on the system used in Illinois for consistency. Each calendar date is assigned a three-digit number, January 1 is 001, January 17 is 017, and so on. In this system, however, we assume every month has 31 days. We know this is not true, but it helps to have a code that allows for leap day. If every month is assigned 31 days (some will never be used) we find that April 10 is 103; three months of 31 days plus the 10 days in April. There will be a total of 372 "days" in this system. This will give us the day and month of the birthdate of the individual. If these numbers are assigned for the male population and adding 600 to the number is assigned to the female population, then we have a system for coding the birthdate and sex of everyone into a single three-digit number. In order to place the year of birth, we place the last two digits of the year in front of the date of birth, with a dash separating the digits. That is, a male born on October 25, 1927 will have 2-7304 as part of his identification number and a female born on October 20, 1932 will have 3-2899 as part of hers. The state of Florida uses a similar system except that each month is assumed to have 40 days and you add 500 for females. For example, a Florida driver license showing 8-0460 would be for a man born on December 20, 1980.

VEHICLE IDENTIFICATION NUMBERS (VIN)

A vehicle identification number (VIN) is a seventeen-digit code using both numbers and letters to identify many things about the vehicle itself. In North America, we have two different types of VINs based on the number of cars manufactured by a specific company each year. The European Union also has two types of VINs and the International Organization for Standardization also has a type of code. They are all closely related, but we will focus on cars manufactured by North American companies that make more than 500 autos per year.

The first three characters are the world manufacturer identifier. The first digit tells where the vehicle was manufactured (1 = U.S., 2 = Canada, 3 = Mexico, J = Japan, K = Korea, and many more). The second digit tells the manufacturing company (G = General Motors, F = Ford, B = Dodge, J = Jeep, and so on). The third digit is the vehicle type (4 = passenger car, 5 = bus, 7 = truck, etc.). Characters four through eight will tell the vehicle attributes such as series, body style, restraint code, and engine type. The ninth character is the check digit determined by a weighting system. The tenth character is for the model year, the eleventh character is the plant where it was manufactured, and the remaining digits are the production sequence numbers. Each manufacturer is allowed a slight variation of this scheme as long as all the same data are encoded into the VIN. You can find VIN decoders for most manufacturers online.

For example, the VIN 1FMCU0H99DUD23381 refers to a Ford Escape SUV manufactured in Louisville, Kentucky (U.S.A.) that has a standard braking system, full airbag system, is rated for 0–6,000 pounds, has a 2.0L 4-cylinder engine, and runs on gasoline. In North America, the last five characters must be numeric digits. For ease of reading, certain letters are not allowed in vehicle identification numbers. These letters are O, Q, and I, as they are too closely related to the numbers 0 and 1.

Our main focus will be determining how the check digit is created. First, each alphabet character is assigned a number. The letters A–I are 1–9 (even though I is never used), the letters J–R are also given 1–9, and the letters S–Z are given the numbers 2–9. Notice the third grouping of letters leaves out the number 1. Now that the letters have been assigned a number, we can look at these 17 digits as numeric values. The weighting system for determining the check digit is to multiply by 8, 7, 6, 5, 4, 3, 2, 10, skip the check digit space, 9, 8, 7, 6, 5, 4, 3, and 2.

Returning to our Ford Escape example, we convert the VIN into numeric equivalents and then multiply accordingly. First, the VIN becomes 16434089?44423381, leaving out the check digit. Then we multiply to get

$$8(1) + 7(6) + 6(4) + 5(3) + 4(4) + 3(0) + 2(8) + 10(9)+$$
$$8(4) + 7(4) + 6(2) + 5(3) + 4(3) + 3(8) + 2(1)$$
$$= 8 + 42 + 24 + 15 + 16 + 0 + 16 + 90 + 36 + 32 + 28 +$$
$$12 + 15 + 12 + 24 + 2 = 372$$

We need the missing check digit number to be the remainder when the sum is divided by 11. If that remainder is 10, we use an X in the 9th place. In this example 372/11 = 33 with a remainder of 9. Looking at the original VIN, we see that 9 is the correct check digit.

IN- AND OUT-OF-CLASS PROJECT

1. Research two different calendar systems. Compare and contrast these systems in a three-page paper. Be sure to cite your references appropriately.
2. If you do not already know, determine the day of the week on which you were born. Next, find all occurrences of your birthday on that day of the week to this point in your life. Finally, determine the next year your birthday will fall on that day of the week. Be sure to show your work.

CHAPTER SIX

GEOMETRY

14.

J ack's choice of major as an architect comes from years of appreciating the beauty in buildings. He has always been interested in both drawing and building, so it was natural to want to create his own structures. As he started to study more, he realized that many of the buildings he found most beautiful had the same general shapes integrated in their design.

6.1 List the first 20 Fibonacci numbers.

1		6		11		16	
2		7		12		17	
3		8		13		18	
4		9		14		19	
5		10		15		20	

6.2 Find the ratios of the Fibonacci numbers found in problem 6.1. For your ratios, divide the larger number by the smaller. That is, your first ratio is 1/1 = 1, the second ratio is 2/1 = 2, and so on.

6.3 Do your ratios in problem 6.2 converge (approach) to any single value? If so, what is it?

6.4 What is the Golden Ratio and how is it used in the design of buildings?

6.5 Find a picture of any building, other than what was shown in the text, and outline at least one Golden Rectangle on it. Attach that picture to this assignment.

6.6 Using the instructions given at the end of the chapter, draw your own Golden Rectangle. Attach your drawing to this assignment.

6.7 Find at least three things in your everyday life that have proportions similar to the Golden Ratio. List them here along with a quick sketch of the item.

While looking into Greek architecture, Jack naturally stumbled across the names Euclid and Pythagoras. It's not that these two built all these buildings, but Euclid is known as the Father of Modern Geometry and Pythagoras had a society that studied Geometry (among other things). The name Pythagoras struck a chord in Jack. He was sure he had heard that name before in his math classes.

6.8 Jack is standing 50 feet from the base of a 120-foot wall. What is the distance from Jack to a pigeon perched at the top of the wall?

6.9 Jack is building a patio. He wants to make sure that his corners are right angles in order to have the best support for his railings. How could he use the Pythagorean Theorem to ensure he has right angles?

15.

One day Jack ventured into the school's art gallery. He loved looking at other people's art to gain inspiration. As he was walking through the rooms, Jack couldn't help noticing the crowd gathered around a big, expensive display. He wondered if anyone ever tried to steal the expensive stuff in this gallery so started looking up to try to find surveillance cameras. Encouraged by the art work he has seen, Jack decides to go home and work on what will be his first masterpiece. He plans to design his own art gallery and figure out how to appropriately protect it.

6.10 Draw a possible gallery with only exterior walls. All walls must be straight and meet another wall at a corner (any angle corner). There will be NO floating interior walls. Your gallery must have at least 13 vertices and not be boring. Use some creativity here, please.

6.11 On a copy of your image from question 6.10, use the process of triangulation, showing the tri-colored image. You may color vertices or edges to show your triangulation. Attach this copy to your assignment.

6.12 On another copy of your image from question 6.10, calculate the greatest number of cameras needed to watch your entire gallery. Then, place as many cameras as you actually need in the positions they should be. Remember, the cameras can sweep left and right in a total circle, but cannot move from their set position. Attach this copy to your assignment.

Now that Jack understands how to lay out a floor plan and how to guard the contents, he wants to construct his first building. He has always been fascinated by unique structures rather than just rectangular buildings. Though he doesn't want some standard building, he does want to make sure it is an appealing look. This leads him back to Plato and the Platonic Solids.

6.13 You can find nets for the Platonic solids at the end of this chapter. Cut out each net along the solid lines. Next, fold along the dotted lines and tape or glue the net together to form the solid. Be sure to label each one. (It may be easier to label while it is still in net form and not in full 3D form.)

6.14 For each of the regular solids, take the number of vertices, subtract the number of edges, and add the number of faces. For each regular solid, what do you get? Use the table to show your results clearly.

Solid	Vertices	Edges	Faces	V – E + F
Tetrahedron				
Cube				
Octahedron				
Dodecahedron				
Icosahedron				

6.15 For each regular solid, imagine slicing off a vertex. What shape is the boundary of the cut? For example, slicing a vertex off a tetrahedron gives a triangular cut. You do the other four.

16.

In order to advertise his new art gallery, Jack wants to design a billboard. He knows that his unusual design choices would attract everyone if he could just get them to look at it. He walked halfway down the block and took a picture to place on the billboard itself. That was when he noticed from this view it looked like his building vanished to the horizon. He started thinking about perspective drawing and decided that would be a much better image than some photograph.

6.16 Use one-point perspective to draw the street in front of Jack's art gallery. Be sure to include building and street details. Do this drawing on a standard sheet of 8.5" × 11" printer paper and attach to this assignment.

6.17 Use two-point perspective to draw the corner view of Jack's art gallery. Be sure to include details. Do this drawing on a standard sheet of 8.5" × 11" printer paper and attach to this assignment.

Satisfied with his initial design, Jack moves on to another project. He loves the concept of decorating the outside walls of his buildings as much as he wants the inside to be beautiful and awe-inspiring. He feels that the modern architecture of "just a wall" leaves little for the area surrounding it. Jack decides to look more into repeating patterns that he could use and discovers the world of tessellations.

6.18 Create a tessellation using an equilateral triangle. Your design must completely cover one sheet of standard printer paper. Instructions for creating tessellations can be found online in many places. Decorate and/or color your tessellations for a more attractive design. Do not duplicate the work of another student. Attach your creation to this assignment.

6.19 Create a tessellation using a square. Your design must completely cover one sheet of standard paper. Instructions for creating tessellations can be found online in many places. Decorate and/or color your tessellations for a more attractive design. Do not duplicate the work of another student. Attach your creation to this assignment.

6.20 Create a tessellation using a hexagon. Your design must completely cover one sheet of standard paper. Instructions for creating tessellations can be found online in many places. Decorate and/or color your tessellations for a more attractive design. Do not duplicate the work of another student. Attach your creation to this assignment.

As an architect, Jack frequently builds scale models of his creations for presentations and public displays. In order to sell a client on a concept, they first need to see what the building will look like in the end. Jack decides to use his skills to create a scale model of his own art gallery.

6.21 Using your design from problem 6.10, build a scale model of your art gallery. You may leave the top open to show your camera placement and tessellation installations, or you may put a roof on it to show the outside building. Either way this should be a completed art gallery with doors, windows, a similar floor plan and exterior decorations from earlier exercises. Include a one-page description of your choices of scaling (include what you expect the original wall heights and lengths to be) along with decoration choices. Write this as a presentation for a client.

14.

GEOMETRY BASICS

Some geometric objects are just math ideas that do not actually exist in the real world. We can visualize them in our heads, but any physical representation will fall far short of the actual object. A geometric point is imagined to have zero size. When you draw a dot on a page, it obviously has size. We imagine it to have size zero. A geometric line is formed by connecting two points along the shortest possible path and extending infinitely in both directions with no thickness. The limiting resources of pens, pencils, and chalk will always give us a line with thickness physically, however. How could we possibly represent a line that is infinite in our world? We could never stop drawing that line! Can you imagine how boring that math lecture would be? We usually work with line segments, or pieces of lines, instead. A geometric plane is a perfectly flat surface that has infinite length and width, but no thickness. Think of it as a sheet of paper that goes on forever in all four directions, yet has no thickness. The idea of continuing forever, or infinitely, is a big concept in math. Yet, we have no way of replicating it in the real world. Our minds can imagine things our bodies could not fathom.

The dimension of an object can be thought of as the number of independent directions in which you could move if you were the object. At a point you could move nowhere, therefore it has dimension zero. On a line you could walk forward (positive) or backward (negative) but really only one direction, therefore one dimension. A line, such as a number line, requires only one coordinate to know where you are. Therefore we see once again that it is one-dimensional. On a plane you could walk forward/backward or left/right, so it has two dimensions. A plane such as the rectangular coordinate system requires both x and y to know a location. Hence we see once again that a plane is two-dimensional. In a three-dimensional space, you can move in three independent directions: forward/backward, left/right, and up/down. Therefore three coordinates are needed to locate a point in three-dimensional space.

Plane geometry is the geometry of two-dimensional objects. We will use circles and polygons as examples. All points on a circle are located at the same distance, the radius, from the circle's center. The diameter of a circle is twice its radius, which means that it is the distance across a circle on a line passing through the center. In math, we use the word "line" to mean a straight line. A line is the shortest distance between two points; therefore, we will use the term "curve" for any line that is not straight. A polygon is any closed shape in the plane made from straight line segments. The word "poly" comes from the Greek meaning "many." Therefore a polygon is a many-sided figure. A regular polygon is a polygon in which all the sides have the same length and all interior angles are equal.

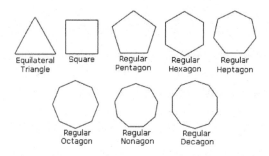

Figure 6A: The first few regular polygons.

The intersection of two lines or line segments forms an angle. The point of intersection is called the vertex. There are three main types of angles: a right angle measures 90 degrees, an acute angle measures less than 90 degrees, and an obtuse angle measures more than 90 degrees. Another type of angle is a straight angle, it is any angle is formed by a straight line and measures 180 degrees.

BY SIDE | BY ANGLE

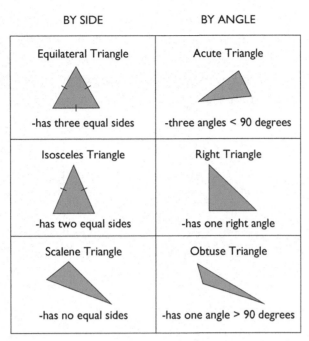

Equilateral Triangle	**Acute Triangle**
-has three equal sides	-three angles < 90 degrees
Isosceles Triangle	**Right Triangle**
-has two equal sides	-has one right angle
Scalene Triangle	**Obtuse Triangle**
-has no equal sides	-has one angle > 90 degrees

Figure 6B: Classifications of triangles by side and by angle.

The perimeter of a plane object is simply the length of its boundary. To find the perimeter of any polygon, add the lengths of all the individual edges. The perimeter of a circle, called the circumference, is related to its diameter or radius by the universal constant π (pi). Ancient people recognized that the circumference of any circle is proportional to its radius. Archimedes' method of discovery included inscribing figures in a circle and circumscribing that same figure around the circle, as shown in Figure 6C. As the figure took on a shape closer to a circle, he was able to make a better estimate of the circumference. Using this method, Archimedes was able to estimate pi to be 3.14. We now know that pi is an irrational number, a number whose decimal representation never ends and never repeats.

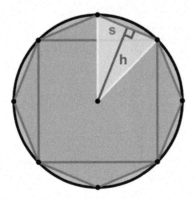

Figure 6C: Archimedes' method of inscribing polygons to estimate the value of ·

The area of a plane object is the amount of space it takes up in two dimensions. Area is frequently used when covering floors with carpet or tiling, or in painting walls or using wallpaper. We think of area as the number of squares of a unit size needed to cover a space. For example, if the wall in your shower is 8' x 5' it would have an area of 40 square feet. This means that if you covered the wall with one-foot tiles, you would need to purchase 40 of them.

NAME	FIGURE	AREA	PERIMETER CIRCUMFERENCE
TRIANGLE	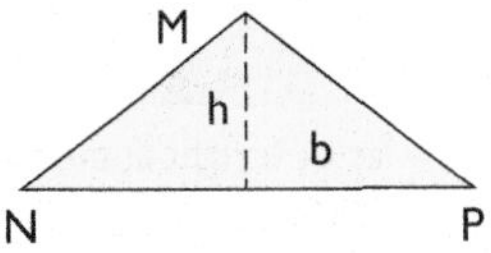	$A = \dfrac{b \times h}{2}$	$P = MN + NP + PM$
PARALLELO-GRAM	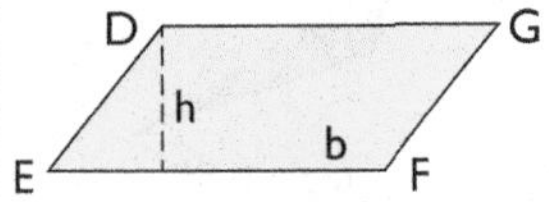	$A = b \times h$	$F = DE + EF + FG + GD$
RHOMBUS	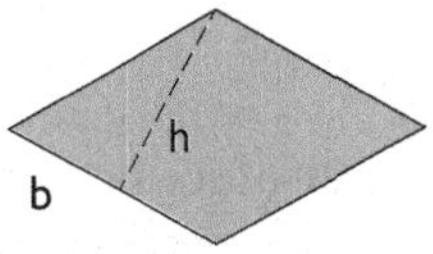	$A = b \times h$	$P = b + b + b + b$ $P = 4b$
RECTANGLE	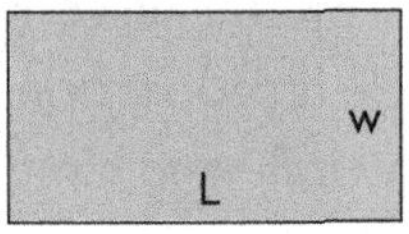	$A = L \times w$	$P = L + w + L + w$ $P = 2L + 2w$
SQUARE	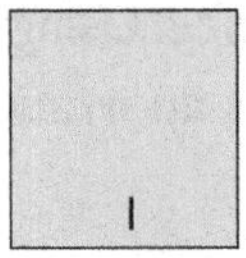	$A = l^2$	$P = l + l + l + l$ $P = 4l$
TRAPEZOID	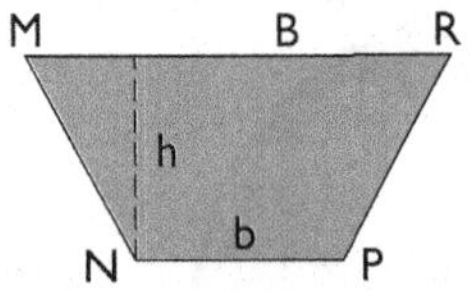	$A = \dfrac{(B \times b) \times h}{2}$	$P = MN + NP + PR + RM$
CIRCLE	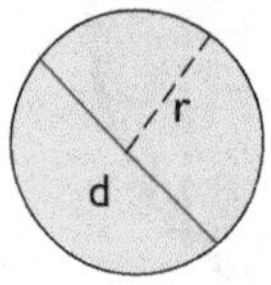	$A = \pi r^2$	$C = 2\pi r = \pi d$

Figure 6D: Some basic shapes with area and perimeter.

Mathematics of three dimensions is the mathematics of our real world. Two of the most important properties of a three-dimensional object are its volume and surface area. In general, volume is the amount an object can hold. For example, my water bottle has a volume of one liter. A box might have a volume of 3 × 3 × 4 = 36 cubic inches. The surface area is the amount of space the surface of the object takes up. The box described would have a surface area of $SA = 2lw + 2lh + 2wh$, which is 66 square inches.

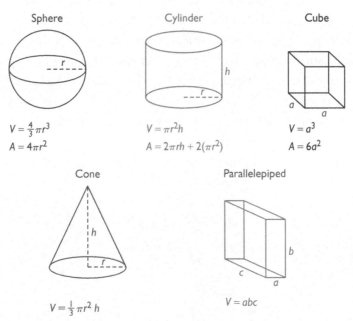

Figure 6E: The volume and surface area of five basic shapes.

Fibonacci

Patterns exist throughout the world. Nearly every pine cone has spirals with eight in one direction and thirteen in the other. Nearly every daisy has thirty-four petals, which is something to keep in mind when playing "they love me, they love me not." Nearly every sunflower has a spiral pattern of twenty-one in one direction and thirty-four in the other. Pineapples have three sets of spirals, most with five, eight, and thirteen spirals. Looking at the growth of many plants, in the segments of the stems, you will find a pattern of 1, 1, 2, 3, 5, 8, 13, 21, 34, 55 and so on. The pattern consists of adding the previous two results to get the next. This is known as the Fibonacci sequence. Is it just a coincidence that the spirals on these natural occurrences fall in with the Fibonacci sequence?

You could continue this pattern until you were too tired to continue. This is what is known as a recurring pattern. The recurring formula for the Fibonacci sequence is that you start with two numbers: one and one. Add these two values to get the next number in the sequence, two. Add the last two values to get the next number, three. Continue forever.

THE GOLDEN RATIO

The Golden Ratio is a value represented by the Greek letter φ (phi). In a Golden Rectangle, divide the length of the base by the height and you will get $\varphi = \dfrac{b}{h} = \dfrac{1+\sqrt{5}}{2} \approx 1.618$. We find the golden rectangle in architecture of both modern and classic times. The Parthenon in Athens, Greece, has many golden rectangles in groupings of the columns, the images above the columns, and even in the original gable of the roofing. See Figure 6F. L'Arc de Triomphe in Paris, France, also has many golden rectangles in its design as shown in Figure 6G. Though we think of the spires and bubble form of the Taj Mahal as its major contribution to architecture, look closely at the columns and window spaces to find many golden rectangles here as well. See Figure 6H. Did the people of these civilizations know about the golden ratio? Did they really plan it out that way with grids, blueprints, and measuring tapes? We know they didn't use our modern tools, but was it all part of the grand scheme? Was it measured exactly to have this pleasing ratio of base to height in so many portions of each structure? We may never know.

Figure 6F: The Parthenon of Athens, Greece.

Figure 6G: The Arc de Triomphe of Paris, France.

Figure 6H: The Taj Mahal in India.

PYTHAGORAS

Pythagoras was a man of many talents. He is known for mathematics, music, and mysticism. His Pythagorean Society is responsible for many of the sounds we hear today in music. They discovered the secrets of the octave and relationships of sounds. The society believed in the power of the pentagram and all that it holds. Mathematically, he is most known for the Pythagorean Theorem. Though this theorem is attributed to Pythagoras, it was known by many different cultures centuries earlier. It is possible that it was named after Pythagoras simply because he was the first to write it for the masses.

The Pythagorean Theorem states: *For any right triangle with side lengths a and b and hypotenuse c, we have $a^2 + b^2 = c^2$.* The converse of this theorem, switching the hypothesis with the conclusion, is also true. It is important to note that this holds only for a right triangle. This is because by definition it is only right triangles that have a hypotenuse; the hypotenuse is the side across from a right angle.

Example 1: On a warm sunny day, Jack decides to fly a kite just to relax. His kite takes off and soars. He lets all 150 feet of the string out and attracts a crowd of onlookers (this is how he met Diane). There is a slight breeze, and a Diane, 90 feet away from Jack, notices that the kite is directly above her. If the string is in a straight line from Jack to the kite, how high is the kite above Diane?

Solution 1: We set up a right triangle with height unknown, let's call it h. The distance from Jack to Diane is 90 feet and the string makes the hypotenuse of 150 feet. We solve for the height using the Pythagorean Theorem:

$$h^2 + 90^2 = 150^2$$
$$h^2 = 150^2 - 90^2$$
$$h^2 = 14400$$
$$h = \pm120$$

Mathematically, the result would have h as plus or minus 120. However, in this real-life situation, we know that the only real possibility is that the kite is 120 feet above Diane, and not 120 feet below her.

15.

THE ART GALLERY THEOREM

The Art Gallery Theorem is a practical application of mathematics in the real world. First, we start with a polygonal closed curve. This means that we have a polygon (many-sided figure consisting of straight lines) in which all edges meet only at endpoints. The corners where the polygonal walls meet will be called vertices. We will not have any floating walls in the middle of the gallery and no walls that extend past the vertices. The Art Gallery Theorem attempts to answer the question of guarding such a polygonal closed curve using only surveillance cameras at vertices that are able to rotate but not move in position. We want to keep the costs down, yet have the entire gallery covered. In order to do this we must consider our lines of sight from each vertex of the gallery.

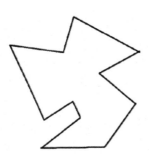

Figure 6I: A possible polygonally closed curve art gallery.

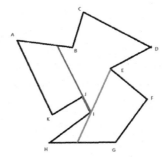

Figure 6J: The same art gallery with lines of sight from vertex E in red and fron vertex I in blue.

In mathematical terms, The Art Gallery Theorem states: *Suppose we have a polygonal closed curve in the plane with v vertices. Then there are v/3 vertices from which it is possible to view every point on the interior of the curve. If v/3 is not an integer, then the number of vertices we need is the biggest integer less than v/3.* How does this translate into more accessible terms? Suppose we have 12 vertices in our art gallery. In this case we would need 12/3 = 4 cameras in order to guard the place. If we have 10 vertices, we need 10/3 = 3.333 … cameras. How could we have 0.333 ... of a camera? This is where the last sentence of the theorem comes into play. Since 10/3 is not an integer, the number of cameras we need is the biggest integer less than 3.33. That is, we would need 3 cameras. The biggest integer function can be thought of as truncation for nonnegative values. Truncation ignores the values to the right of the decimal point. It is important to note that The Art Gallery Theorem tells us how many cameras we will need, but it does not tell us where we should place them.

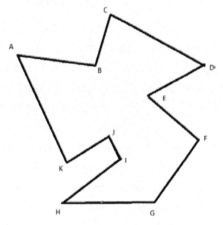

Figure 6K: I would place only 2 guards for this art gallery. Either one at vertex B and another at G, or one at E and another at F. With 11 vertices, the Art Gallery Theorem states we will need at most 3 guards. In this case we can use less.

It is natural to ask why we divide by 3. The three is the most important part of the theorem, as it turns out. The secret to The Art Gallery Theorem is based on triangles. Remember that a triangle is a polygon with three sides (and three angles). If we triangulate our art gallery (connect vertices in such a manner that only triangles remain on the inside) we can color each edge with one of three colors, say red, green, and blue. As you triangulate and color edges, pay attention to the colors, as each triangle in the process must contain all three colors on the edges.

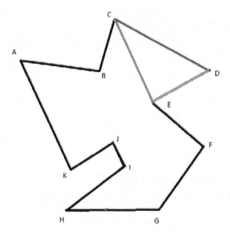

Figure 6L: The first triangle in the triangulation of an art gallery.

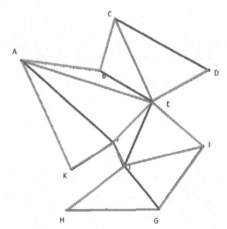

Figure 6M: An art gallery that has been triangulated and tri-colored.

How will triangulating help to place the cameras? Notice that we can now place a camera according to the triangles in order to see everything. One camera for each triangle, then eliminate extra cameras as you find them. Another strategy would be to place a camera then mark the triangles it can view; these will not need another camera. Continue placing cameras until all triangles are covered.

PLATO AND THE PLATONIC SOLIDS

Plato was a Greek astronomer who lived from 427 to 347 B.C.E. He believed that the heavens must have a perfect geometric form, and therefore argued that the sun, the moon, the planets, and the stars must move in perfect circles. This was held to be the common belief until he was finally proved incorrect nearly 2,000 years later. The perfect geometric forms to which he referred are known as perfect solids, or Platonic Solids. Plato was also well known for inventing a moralistic tale about a fictional land called Atlantis.

Objects that are symmetrical look the same from several different views, or two sides are mirror images of each other. Symmetric solids are referred to as regular, or Platonic, solids. If we look first in two-dimensional space, there are many regular polygons. In order to be regular, a polygon must have all sides of equal length and all angles of equal measure. The more sides we add, the closer we will get to a circle. A circle is an infinitely many sided polygon and is the perfect image of symmetry. Some regular polygons with their symmetries are given in Figure 6N.

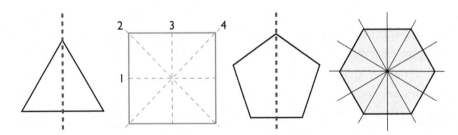

Figure 6N: Symmetries of a triangle, square, pentagon and hexagon..

When we jump to three dimensions, we have only five regular (Platonic) solids. Each of the five Platonic solids are regular because the same number of sides meet at the same angles at each vertex and identical polygons meet at the same angles at each edge. For centuries, the Platonic solids were associated with mystical powers, as many mathematical concepts were at that time. Euclid had written about these solids and the Pythagoreans knew that there were only five and held them in awe. So why are they named after Plato? Most likely they are called the Platonic solids because Plato tried to relate them to the five elements: fire, earth, wood, metal, and water.

It was not only the Greeks who found the allure of the Platonic solids. German mathematician Johannes Kepler tried to use the Platonic solids to describe planetary motion. In his time, there were six known planets. Kepler showed that it is possible to take the five regular solids, put one inside the other, and have the sizes of inscribed and circumscribed spheres about these solids reveal the sizes of the orbits of the planets. Once the seventh planet, Uranus, was discovered Kepler's theory fell apart.

The five Platonic solids are

1. The tetrahedron, made up of four identical equilateral triangles. Each of these triangles is called a face of the tetrahedron. It also has four vertices and six edges. The most common example of a tetrahedron is a pyramid.

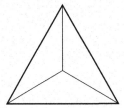

Figure 6O: A tetrahedron.

2. The cube, made up of six identical squares. Each of these squares is called a face of the cube. It has eight vertices and twelve edges. Perhaps the most common of the Platonic solids, the cube is the shape of dice, play blocks, and even a car. (Though the car is stretching it a bit.)

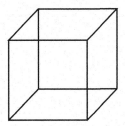

Figure 6P: A cube.

3. The Octahedron, made up of eight identical equilateral triangles. (You were expecting octagons weren't you?) These eight triangles make up the faces of the octahedron. It also has six vertices and twelve edges.

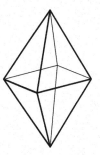

Figure 6Q: An octahedron.

4. The dodecahedron, made up of twelve identical pentagonal faces. This is the only Platonic solid that uses pentagons. The prefix "dodec" gives us do = 2 and dec = 10 for a total of twelve faces. There are twenty vertices and thirty edges! This one is pretty challenging to draw.

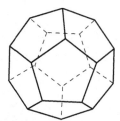

Figure 6R: A dodecahedron.

5. The icosahedron, made up of twenty identical equilateral triangles. Each of these triangles makes up the twenty faces. It also has twelve vertices and thirty edges.

Figure 6S: An icosahedron.

The Platonic solids can be summarized in the table from Figure 6T:

	#Vertices	# Edges	# Faces	# Faces at each vertex	#Sides at each face
Tetrahedron	4	6	4	3	3
Cube	8	12	6	3	4
Octahedron	6	12	8	4	3
Dodecahedron	20	30	12	3	5
Icosahedron	12	30	20	6	3

Figure 6T: Summary of Platonic solids.

Taking a closer look at the table, we can make a few observations. The number of faces of the cube equals the number of vertices of the octahedron and the number of vertices of the cube equals the number of faces of the octahedron. The number of edges of both figures is the same. Is this unusual? Look closer at the dodecahedron and icosahedron. These two platonic solids have a similar relationship. Where does this leave the tetrahedron? From the cube we can construct an octahedron using midpoints of the cube faces. This is called a dual. The process of creating one solid from another is duality. The dual of a cube is an octahedron, the dual of the dodecahedron is the icosahedron, and the tetrahedron is self-dual. Five images are shown in Figure 6U.

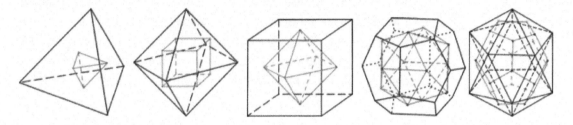

Figure 6U: Platonic solids and their duals.

16.

PERSPECTIVE

Have you ever looked down a long road or train tracks and noticed that it looked like it was getting "thinner" in a way? This is an optical illusion created by perspective. Perspective is defined as the way in which an object appears to the eye. To draw an image in perspective, we need to have a horizon and a vanishing point on that horizon. The horizon is an imaginary line in which the ground "ends" and the sky "begins" in relationship to your image. The vanishing point is an imaginary point in which everything seems to disappear, or vanish.

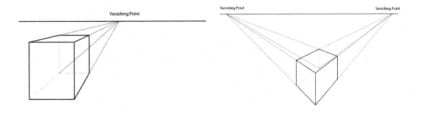

Figure 6V: The one point perspective with horizon line and vanishing point. Two point perspective shows both vanishing points.

 How to draw in perspective.

TESSELLATIONS

The term "symmetry" has many meanings. In *The Last Supper*, by Leonardo da Vinci, symmetry is about balance in that the disciples are grouped in four groups of three, with two groups on each side of the central figure of the Christ.

Figure 6W: *The Last Supper* by Leonardo da Vinci.

A human body has symmetry because a vertical line drawn through the head and navel divides the body into two (nearly) identical parts.

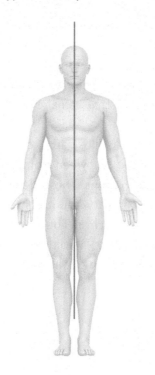

Figure 6X: A human body with line of symmetry.

Symmetry can also refer to a repetition of patterns. This type of symmetry can be found on pottery of Africa, the Aztecs, Greeks, and many others.

Figure 6Y: An African pottery plate, an Aztec terracotta vase with repeating patterns and translation symmetry, and a Maya pottery urn.

There are three main types of symmetries: reflection, rotation, and translation. In reflection symmetry, an object remains unchanged when reflected across a straight line. This can also be referred to as a mirror image. Consider the capital 'A' of the English alphabet. If we draw a line straight through the middle of the letter, we see a mirror image on the right and left. This is reflection symmetry. Rotational symmetry occurs when an object remains unchanged when rotated through some angle about a point. Consider a stop sign. If we rotate a stop sign 45 degrees, we have the same stop sign shape again (even though the word wouldn't show as expected). Translation symmetry is a pattern that remains the same when shifted in a straight line of any direction.

Symmetry is found all over the art world. An illustration of rotational symmetry can be found in Gustave Dore's engraving *The Vision of the Empyrean* from 1870. Victor Vasarely's *Supernovae* started symmetric but then had slight deviations. These deviations can make the work even more powerful. Perhaps the most famous artist when it comes to tessellations is M.C. Escher. Maurits Cornelis Escher (1898–1972) was born in The Netherlands and is one of the most famous graphic artists of all time. He made over 448 lithographs, woodcuts, and wood engravings, along with over 2,000 sketches and drawings. His work can be viewed on this website: http://www.mcescher.com/

Figure 6Z: Gustave Dore's illustration of *The Vision of the Empyrean* (from *The Divine Comedy*) and an example of tiling by Victor Vasarely (right).

A form of art called tiling, or tessellation, involves covering a flat area such as a floor or wall with geometric shapes. Tilings usually have regular, or symmetric, patterns. Specifically, a tessellation is an arrangement of polygons that interlock perfectly with no overlapping. The only three regular polygons that can tessellate a plane are the equilateral triangle, the square, and the hexagon. This is because in order to tessellate we must make our vertices meet without overlap and without any space left over. The formula $A = \dfrac{180(n-2)}{n}$, where n is the number of sides of the polygon, gives us the measure of each vertex of the polygon. For example, the equilateral triangle has angles measuring 180(3–2)/3 = 60. You can fit six equilateral triangles together at a vertex to make a full circle and therefore cover the plane. The square has angles 180(4–2)/4 = 90. You can fit four squares together at a vertex to make a full circle and therefore cover the plane. The hexagon has angles 180(6–2)/6 = 120. It will take three hexagons to make a full circle with no overlap; keep in mind that a circle must encompass a full 360 degrees. If we try this process with a different regular polygon, say the octagon, we find that the angle measure at a vertex is 180(8–2)/8 = 135. However, 135 is not a factor of 360, so no matter how we try, we cannot use an octagon to tessellate a plane. We would either have gaps at the vertices or overlap, neither of which is allowed. We could use a combination of polygons to tile a plane as long as there are no gaps and no overlaps; the angles must add to 360 degrees at each vertex.

SCALING

When you take a picture, you are digitizing the world around you. (Unless you still use film, then you are just cool.) The image that results is smaller than the actual object, generally, but you can use zoom features to change that sizing. This process of changing sizes of an object uniformly is known as scaling.

When constructing a building, contractors will use blueprints. A blueprint is a two-dimensional representation of a three-dimensional structure. If they are building a skyscraper and want to represent it exactly, they would need a great deal of paper. Instead, they use a scale factor in order to render the drawing. Perhaps they will allow ten feet of building to be represented by one inch on the paper. When using plans to build a bookcase, you may have a scale factor of one centimeter on the plans is equivalent to one half of a foot for the bookcase. We frequently see scaling when looking at maps. Take a look at any road map—it should have a legend that tells you so many inches are equal to a certain number of miles.

Scaling can be done in many different ways. You could scale your lengths by a factor of three, your heights by a factor of five, and your depths by a factor of two. Though I'm not sure why you would want to do this, it is possible as long as you clearly state your scale factors. This would change the rendering of your object as the three dimensions would not have the same relations to each other in the model as they do in the object you are trying to model. For this reason, we will use a common scale factor. That is, if you scale a length by a factor of ten (if the original is ten feet, the model will be one foot), then we will scale the other dimensions by ten as well. Using uniform scaling will allow us to take a wall that is 8' high and 10' long, and represent it by a model that may be 8" high and 10" long. Other scale factors are possible and definitely encouraged.

CONSTRUCTING YOUR OWN GOLDEN RECTANGLE

First, draw a square. Not an "it looks close" kind of square, but an actual "I measured it" square. Next, connect the midpoint of the base of the square to the northeast corner of the square with a straight line segment. Extend the base of the square with a straight line segment off to the east, like a landing strip. Now, draw part of a circle whose center is the midpoint of the base and whose radius extends to the northeastern corner of the square; note where the circle portion hits the landing strip. The line segment drawn inside the square from the midpoint to the northeastern corner is actually a radius of the circle arc drawn. Next, construct a line perpendicular to the landing strip and passing through the point where the circle hit the landing strip. Extend the top edge of the square to the right with a straight line until it hits the perpendicular line just drawn. Finally, erase the excess landing strip to the right of the arc.

 How to draw a golden rectangle.

IN- AND OUT-OF-CLASS PROJECTS

1. A scale model has been assigned here in Chapter Six in constructing an art gallery. Choose another item to build a scale model of; it may be a building on campus, your home, your favorite building, or other object. Include a one-page paper detailing your choices in scaling and building this model.
2. Write a five-page research paper on one of the many mathematicians mentioned in this chapter. Be sure to properly cite your references.

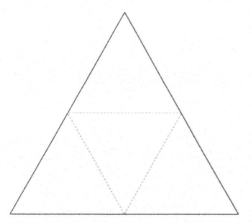

Figure 6AA:Tetrahedron net. Cut around the outer edges and fold on the dotted lines.

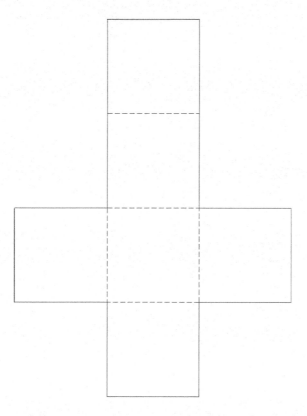

Figure 6AB: Cube net. Cut around the outer edges and fold on the dotted lines.

Figure 6AC: Octahedron net. Cut around the outer edges and fold on the dotted lines.

Figure 6AD: Dodecahedron net. Cut around the outer edges and fold on the dotted lines.

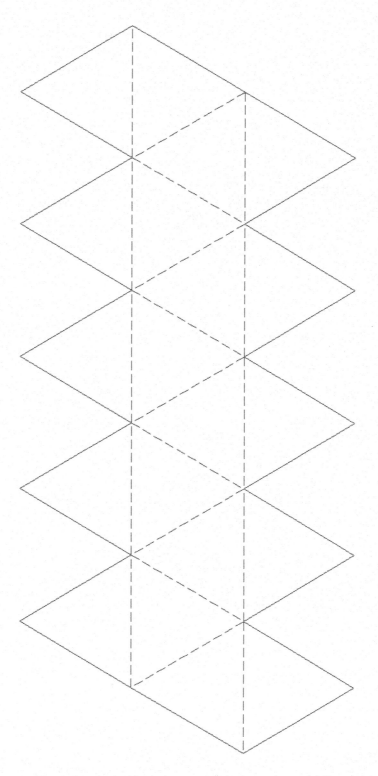

Figure 6AE: Icosahedron net. Cut around the outer edges and fold on the dotted lines.

CHAPTER SEVEN

CIRCUITS, PATHS, AND SCHEDULES

17.

Jack's dad has always taught him that money is something to be earned and that only through earning your own money can you truly appreciate the items you buy. In order to start earning money, Jack decides to take a job as a paperboy. He has to wake early to deliver papers before school. Some days, Jack has club meetings after school. He tries to keep on top of everything and complete all his tasks in as little time and with as little wasted effort as possible.

7.1 On Jack's first day as a paperboy, he noticed that not many cars were out that early. He decided that he could ride his bike down the middle of the street (don't try this at home) and toss papers to houses on both sides of the street at the same time in order to save effort. His boss gave him a simple route for his first day. Jack travelled it as shown. The streets are shown with edges and intersections are the vertices. After receiving calls from irate customers, his boss was upset with this choice of route. Why?

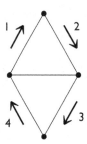

7.2 On Jack's second day as a paperboy, he listened to what his boss had told him and decided to take a different approach to the same route, as shown. This time, Jack wasn't very happy with his choice of route. Why?

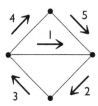

7.3 If Jack's goal was to ride down each street of the route given exactly once, was it a reasonable goal? Explain your answer.

7.4 As part of his club fundraising activity, Jack goes door to door selling popcorn. He needs to travel each edge of his paper route twice, one for each side of the street. Draw a graph that may show his path.

7.5 Is the route you found in 7.4 the most efficient? Can you find a better route? How do you know if it is most efficient? Explain your answers using complete sentences.

As a high school senior, Jack has been looking forward to touring local college campuses so he can dream of what it will be like when he finally gets there. He daydreams through most of the tour, checking out the "views" on campus, before arriving at the union square. The tour guide keeps talking about studying, camaraderie, and some other stuff Jack isn't paying attention to when he notices the sidewalks of the square.

7.6 For the sidewalks of the union square shown below, show how Jack could walk every sidewalk at least once in a tour that starts and ends at his current location marked by the letter A. Sidewalks are shown as edges and intersections of sidewalks are given as vertices.

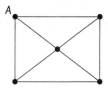

18.

During spring break of his senior year of high school, Jack takes a road trip with a few of his friends. They decide to follow their favorite band, Somebody's Weird, on a part of their tour through the cities of Phoenix, AZ; Santa Fe, NM; Salt Lake City, UT; and Denver, CO.

7.7 Make a table of values showing the distances between each pair of cities. Represent this information by drawing a weighted complete graph with four vertices as well.

	Phoenix	Santa Fe	Salt Lake City	Denver
Phoenix				
Santa Fe				
Salt Lake City				
Denver				

7.8 Use the weighted graph you found in 7.7 to find the distance of the three distinct Hamiltonian circuits in the graph. List these circuits starting at Phoenix, AZ, as that is closest to their home.

7.9 Which of the three circuits gave the least distance traveled? Why would this matter to the band?

7.10 If Jack applies the nearest-neighbor method starting in Phoenix, what circuit would be obtained? Does the answer change if he were to start in Santa Fe? Why or why not?

7.11 If Jack used the sorted-edges method, what circuit would be obtained? Is this the optimal answer?

In the end, Jack was just entertaining himself with possible trips. They were following their favorite band! They would drive as far as they needed and in whatever order the concerts were scheduled to see that many shows in a week. Graduation day has finally come and Jack convinces his parents to let him host an amazing party. Though Jack wants all of his friends to attend, he realizes that they do not all have driver licenses and so he promises to give them all a ride home at the end of the night.

7.12 After the party, Jack agrees to drive Josh, Andrew, and Noah home. If the times, in minutes, to drive between his friends' houses are shown in the figure below, what route gets Jack back home the quickest?

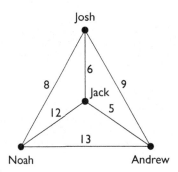

7.13 If Josh had the earliest curfew so had to be home first, what route should Jack follow in order to get back home as quickly as possible?

7.14 Rather than heading back home, Jack is going to drive his friends home and then stay with Andrew. What would be the fastest route?

19.

For his last summer at home, Jack offers to act as his dad's assistant whenever he needs one. His first job with his dad is adding on a room to their house.

7.15 Jack is given the following tasks: (a) Erect sidewalls, (b) erect roof, (c) install plumbing, (d) install electric wiring, (e) install lighting in the ceiling, (f) install wall air conditioner, (g) lay foundation, (h) lay tile flooring, (i) obtain building permits, and (j) put in door that adjoins new room to the existing house. Determine reasonable time estimates for these tasks and a reasonable order-requirement digraph. What is the fastest time in which these tasks can be completed?

7.16 Jack's second job with dad involves setting up the trampoline that his little brother is getting for his birthday. There are eight steps involved with setting up this trampoline: Step 1—Remove all parts from the box, Step 2—Connect the frame including legs, Step 3—Attach surface to the frame with springs, Step 4—Attach safety foam to the edge, Step 5—Erect frame for safety net, Step 6—Attach snap buckles of net to the top frame, Step 7—Tie cord of the net to trampoline main frame, Step 8—Place warning stickers and post family rules. Give reasonable time estimates for each step and construct a reasonable order-requirement digraph. What is the least amount of time it will take to set up this trampoline?

7.17 Jack's last home improvement project with his dad is a kitchen remodel. They will need to clean the kitchen, scrape the walls to remove all the old paint and tile, prime the walls, install wallpaper, scrape the paint off the ceiling, paint the ceiling, replace the old floor with new floor tiles, install a new stove, install a new sink, install new cabinets, install a new countertop, and install a new refrigerator. Draw an order-requirement digraph for these tasks, giving reasonable estimates for the times to do each task. Find a critical path for your order-requirement digraph.

Having spent a few months away at college, Jack is looking forward to his trip home for Thanksgiving. He loves and misses his mom's cooking, but her turkey is the best thing he has ever tasted. One of his favorite parts of Thanksgiving is that the entire family pitches in to help cook and clean so everyone gets to watch football, too.

7.18 Suppose Jack, his parents, and his two younger siblings (ages 7 and 12) are preparing a Thanksgiving meal for ten people. List the tasks that must be completed and the types of processors that are involved. Can any of these tasks be done simultaneously?

7.19 Before the NFL games on Sunday, Jack's family tries to quickly prepare a meal with tasks represented by the order-requirement digraph given. If Jack prepares the meal alone, how long will it take?

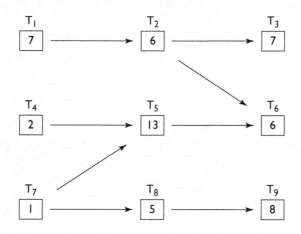

7.20 If Jack is helped by his mom, how long will it take them if the tasks are scheduled using the list T5, T9, T1, T3, T2, T6, T8, T4, and T7 and the list-processing algorithm still using the order-requirement digraph of exercise 7.19.

7.21 If Jack can talk both of his parents into helping him, how long will it take if the tasks are scheduled the same as they were in 7.20?

7.22 What would be a reasonable set of criteria for choosing a priority list in this situation?

20.

Back at college, Jack returns to his work-study job in the copy center. With finals quickly approaching, they have a lot of work to accomplish quickly. The times for the different sets of independent documents that need to be copied (in minutes) are 12, 23, 32, 13, 24, 45, 34, 34, 14, 21, 34, 53, 18, 63, 47, 25, 74, 23, 42, 42, 16, 16, and 76.

7.23 Jack is asked to construct a schedule using the list-processing algorithm with the three machines they currently have. Show that schedule.

7.24 Jack's boss wants to know how much it would help if they were able to rent a fourth machine. Construct a schedule using the list-processing algorithm with four machines.

7.25 Use the decreasing-time-list algorithm to create schedules for both three machines and for four machines.

7.26 The copy machines the copy center use are older than what they would like. In order to keep them running efficiently, each copier requires an 8-minute rest period for any copy task over 45 minutes. Use the decreasing-time-list algorithm—with the given times modified to take into account the overheating of the machines—to schedule the tasks on three machines.

7.27 Several students have brought their term papers to be copied. Suppose it takes 4 seconds to copy one page. Term papers of 10, 8, 15, 24, 22, 24, 20, 14, 9, 12, 16, 30, 15, and 16 pages are to be photocopied. How many copy machines would be required using the first-fit-decreasing algorithm, to guarantee that all term papers are photocopied in two minutes or less? Would your answer be different if worst-fit decreasing were used instead?

17.

NETWORKING BASICS AND EULER CIRCUITS

The study of networks is a study of time- and money-saving measures. The roots of street networks, and therefore operations research, can be traced to something called the Königsberg Bridge Problem. In 1735 Leonhard Euler proved that it was impossible to travel each of the seven bridges of Königsberg exactly once while ending your journey at the same location you started. In order to better understand this problem, and its significance, we need some basic terminology.

Figure 7F: The seven bridges of Königsberg in bridge form and drawn as a graph.

In networking, a graph refers to a mathematical structure consisting of points and lines. The points are called the vertices of the graph and can represent cities, intersections, campus buildings, anything of interest. The lines, called edges, connect the vertices, showing a particular relationship. In the Königsberg bridge problem, we could represent each island as a vertex and each bridge as an edge, thereby showing which bridges connect specific islands, or parts of the city. A path is a connected sequence of edges in a graph. Think of it as your Mapquest directions for this particular graph. The path that Euler was interested in was a circuit. A circuit is a path that starts and ends at the same vertex. Due to his work on this problem, a special type of circuit was named after Euler. In an

Euler circuit you start and end at the same vertex, traveling each edge of the graph exactly once, exactly as he tried to do with the bridges.

We're still not quite ready to solve the bridge problem. How did Euler know it couldn't be done? Let's take a closer look at each vertex. We say the valence of a vertex is the number of edges connected to that vertex. For example, if you consider the intersection of 5th Avenue and Broadway in New York City as your vertex, it has valence four. Broadway extends from this intersection in both directions as does 5th Avenue. If you consider the intersection of Santa Monica Boulevard and Ocean Avenue in Los Angeles, your valence is now three, as this is where Santa Monica Blvd ends. The last term we need makes sure there is a relationship between each of the vertices. A graph is said to be connected if for every pair of vertices there is at least one path connecting the two vertices. In other words, a graph is connected if you can get from any vertex of the graph to any other vertex in the graph, even if you have to go through several vertices on your way.

We now have everything we need to see Euler's solution. He was able to prove that if a graph is connected and has all valences even, then an Euler circuit exists. The converse of this is also true; that is, if an Euler circuit exists, then the graph must be connected and have all valences even. This theorem, Euler's Theorem, allows us to look at a graph and know if an Euler circuit exists. It does not, however, tell us how to find that circuit. If a certain graph has some odd valences, we can use a method known as Eulerizing in order to make it an Euler circuit. To do this we add as few edges as necessary to make all valences even. This will result in the crossing of a specific bridge twice, or some type of backtracking in your path, but will make an Euler circuit possible so that you arrive back at where you started your path.

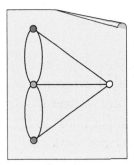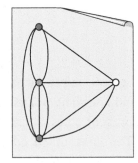

Figure 7G: The seven bridges problem in original graph form and the solution after Eulerizing the graph.

When we Eulerize a graph, we don't want to randomly add edges. The Chinese Postman Problem, studied in 1962, asserts that adding the fewest edges possible will help to minimize the length of the circuit. Our routes, or paths, will get more complicated if we add in that each edge must be traveled twice. What practical applications could this have? A garbage collector needs to make a pass down each side of the same street for residential collection. A snow plow operator needs to plow both sides of a street, traveling in the appropriate direction with each plow. Graph theory, especially circuits, has many applications in our everyday lives. When these situations arise we may use a directed graph, or digraph. A digraph is a graph in which each edge has an arrow indicating the direction of the edge. This is most important for one-way streets, but possibly a street has houses on only one side that a postal carrier needs to deliver mail to.

 Eulerizations.

HAMILTONIAN CIRCUITS

Irish mathematician W.R. Hamilton was one of the first to study a different aspect of graphs where the focus was more on visiting vertices rather than traveling edges, as was Euler's focus. A Hamiltonian circuit is a circuit, so you must end where you begin, but it has the added feature that you must visit each vertex once and only once. Perhaps you have errands to run on your day off. You need to go to a doctor's appointment, make a deposit at the bank, get some groceries, pick up dry cleaning, and pick up a prescription from the pharmacy before returning home. You don't want to stop at the dry cleaners, or even waste gas by driving by it, multiple times if you can avoid it. A Hamiltonian circuit can provide the most efficient way for you to accomplish your errands.

For Hamiltonian circuits, we will use weighted complete graphs. A weight is a number added to an edge that could be the cost of running electricity, miles between cities, or even length of time an activity will take. A complete graph is one in which every pair of vertices is joined by an edge. Keep in mind this is different than a connected graph. In a connected graph you can get to each vertex eventually. In a complete graph you can get to each vertex directly. There may be many Hamiltonian circuits for any particular weighted complete graph. How can we find which Hamiltonian circuit has minimum cost? Unfortunately this is a brute force method: First, list all possible Hamiltonian circuits from your starting vertex. Second, add the distances of each path. Third, choose the path of least value. As we are concerned with vertices here rather than edges, each Hamiltonian circuit will usually have a different cost associated with it. Steps two and three are pretty easy as far as arithmetic is concerned. The problem is with step one and finding all possible Hamiltonian circuits.

In order to find these circuits it may be best to take an organized approach. Suppose we are starting at home and wish to go to the bookstore, the bank, the grocery store, and then home again, not necessarily in that order. We can use the method of trees to help us make our Hamiltonian circuits.

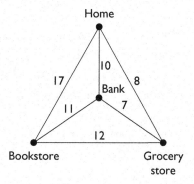

Figure 7H: A weighted complete graph of running errands.

We start with the vertex of home. We then have three branches taking us to either the bookstore, the bank, or to get groceries. Suppose we choose to go to the bank first, from that branch we would now branch off again to either the bookstore or grocery store, as these are the only vertices remaining. Keeping in mind that this is a complete graph, and we can get to any vertex from any other vertex, this allows us to write all possible Hamiltonian circuits as shown. We now follow each branch down, adding in distances in order to complete the process.

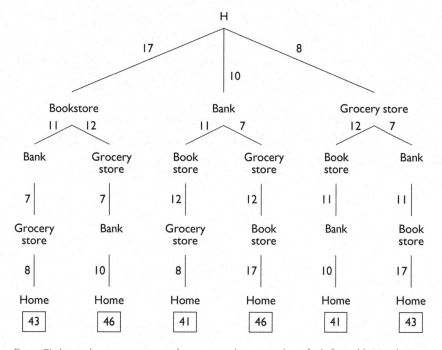

Figure 7I: A tree diagram starting and returning to home in order to find all possible Hamiltonian circuits.

Hamiltonian circuits led to the Traveling Salesman Problem (TSP). This problem is aptly named as it is trying to get a traveling salesman through his route with as little cost as possible. There are two ways we will examine for solving the TSP; the first is called the nearest-neighbor algorithm. An algorithm is just a step-by-step procedure. In the nearest-neighbor algorithm, from each vertex you select the route that is the shortest (or cheapest, in general of least weight). By always choosing the shortest edge, or nearest neighbor, you make your way to each vertex of the weighted complete graph, as shown in Figure 7J. It seems that this would be our optimal, or best, route. However, this is an example of a greedy algorithm and will sometimes leave you in a bad spot in which you must take the most costly route on your return.

 Traveling Salesman Problem visualization.

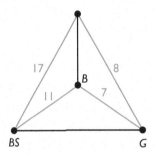

Figure 7J: The nearest neighbor algorithm.

Another option is the sorted-edges algorithm. We make a list of the weights of the edges in an increasing order, low to high. We select each edge by keeping two things in mind: First, we don't have to make the circuit as we go. We can add edges throughout the graph as long as they end up in a circuit. Second, we cannot add an edge that would prevent a Hamiltonian circuit from being formed. This means we cannot use three edges from a particular vertex and we cannot close off a path that leaves out a vertex. Again, this algorithm does not necessarily lead to an optimal solution. Sometimes we need to accept that optimal might not be guaranteed, but the method gives a "good enough" solution to our problem.

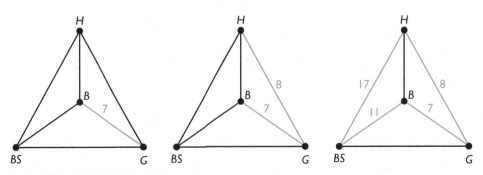

Figure 7K: The sorted edges algorithm in three steps.

19.

CRITICAL PATH ANALYSIS

Running errands allows us to choose which order we visit each vertex. Sometimes we must do things in a specific order. For example, after attending class in the morning, you have a doctor's appointment in the afternoon. Only after the doctor's appointment can you go to the pharmacy for the medicine that has been prescribed. Only after picking up your medication can you take it as prescribed once you return home. When certain tasks cannot be done in a random order we use what is called an order-requirement digraph.

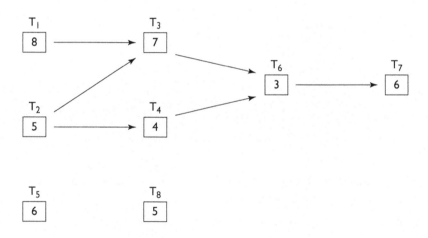

Figure 7L: An order requirement digraph. Tasks 5 and 8 are called independent tasks as they are not required to be completed in any particular order.

In order to estimate the time required for this example, we would have to have time estimates for each task and then choose the path of the longest time. This seems counterintuitive at first, especially after our dealings with circuits always wanting the shortest path. We choose the longest path as the critical path because the longest path is the shortest time in which everything can get accomplished.

Let's take an everyday action and break it down into tasks. In order to eat a peanut butter and jelly sandwich you need to conduct the following tasks: T1—Grab peanut butter from the cabinet and open the jar, T2—Grab jelly from the refrigerator and open the container, T3—Grab loaf of bread and remove two slices, T4—Grab two knives from the drawer, T5—Spread peanut butter on one slice of bread, T6—Spread jelly on one slice of bread, T7—Close the sandwich, T8—Close peanut butter container and put it away, T9—Close jelly container and put it away, T10—Enjoy your sandwich. You probably didn't know making a pb&j was so labor intensive did you?! We can estimate times for each of these activities, in seconds, and place them in an order-requirement digraph as shown.

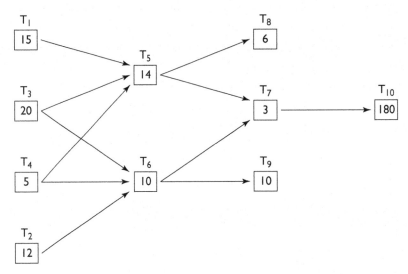

Figure 7M: An order requirement digraph for making a peanut butter and jelly sandwich.

The critical path for making and eating this peanut butter and jelly sandwich would have time 217 seconds. If you had enough people helping you with the process so that your only focus was to work on the longest task line and your friends did tasks not on that line for you, you could make and eat your sandwich in less than four minutes.

SCHEDULING TASKS

The people who would be helping to make your peanut butter and jelly sandwich could be thought of as processors, or machines. Each processor is scheduled a task. Once that task is complete, they move on to the next available task. In order to keep each processor, or person, busy with as little idle time as possible, we need to find the most effective way to schedule these tasks in accordance with the order-requirement digraph of a particular situation. In order to streamline our discussion and understanding, we will take a few things for granted. First, once a processor begins a task it will continue that task without interruption until the task is completed. Second, a processor cannot opt to skip a task that is available and remain idle. (Your friend with a peanut allergy shouldn't even be in the house if you're eating a peanut butter sandwich so he can't use his allergy as an excuse for not getting the peanut butter ready.) Third, we have to abide by the order-requirement digraph. Our last assumption is that we will also have a priority list. Is it a priority for you to put the peanut butter and jelly away before eating the sandwich, or can you wait to clean it all up after?

In order to appropriately schedule tasks with our processors, we will use the list-processing algorithm. This algorithm is not designed to guarantee the optimal solution, but it is a viable algorithm. To use the list-processing algorithm, we will need to number our processors, have a priority list, and have an order-requirement digraph. We will assign the first ready task on the priority list that has not already been assigned to the lowest-number processor that is not currently working on a task. Suppose we wish to schedule the peanut butter and jelly sandwich with two people (processors), you and a friend. You will be the first processor and your friend will be processor two. Our priority list will be T1, T2, T3, T4, T5, T6, T8, T9, T7, and T10. We use the list-processing algorithm to arrive at the schedule shown in Figure 8N.

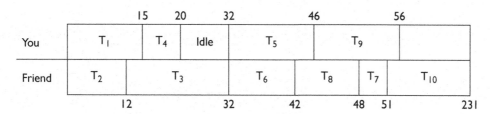

Figure 7N: The list processing algorithm to schedule two processors in making a peanut butter and jelly sandwich.

Notice that this is definitely not an optimal solution, as your friend just ate your sandwich! As we look for the next available task, we may have to skip some on the priority list because they are not yet ready. For this reason it is important to always start from the beginning of the list to make sure you pick up any tasks that had been previously skipped.

20.

INDEPENDENT TASKS

Not every set of tasks that we need to do have an order to them. Suppose you have readings for four classes over the weekend; you do not need to finish the readings in any particular order because the classes are not related. Tasks that are not related to each other, and therefore do not have an order-requirement digraph, are said to be independent tasks. As our example, you and your roommate decided to clean up your house for a mid-semester party. You need to clean the main bathroom, 30 minutes, clean the kitchen, 25 minutes, vacuum, 35 minutes, dust, 15 minutes, shop for supplies, 50 minutes, and clean the backyard, 90 minutes.

In order to organize and schedule these tasks, we could use the decreasing-time-list algorithm. The name is self-explanatory as we will list all the completion times in a decreasing list with longest time first and shortest time last: 90, 50, 35, 30, 25, and 15. We can now schedule this task list with you and your roommate as the two processors.

		90	115	130
You	Yard		Kitchen	Dust
Roomate	Shop	Vacuum	Bathroom	Idle
	50	85	115	

Figure 70: A decreasing time list schedule for you and your roommate to get ready for a party.

There are other decreasing-time-list approaches for more processors. Suppose you have eight shelves for your books. Each shelf can hold fifteen books. You have your books divided into genre with the following numbers of books per section: 12, 12, 11, 10, 9, 6, 5, 5, 5, 4, 4, 3, 3, 3, 2, 2, and 1. Our goal is to fill as many shelves as possible before using more shelves. This is what is known as a bin-packing problem. We want to fill a bin, or bookshelf, with as little empty space as possible. There are several methods of solving this problem, including next-fit-decreasing, first-fit-decreasing, and worst-fit-decreasing algorithms.

The next-fit-decreasing algorithm tells us to use our decreasing-time list, or in our case a decreasing-size list, and place the first item on the first shelf. If the second item fits on that shelf, place it; otherwise the first shelf is now complete and we place the second item on the second shelf. We can never go back to a previous shelf—we can use only the current shelf or the next shelf. This may leave open spots along the way. For this example, next-fit-decreasing would require us to use all eight shelves.

	15
12	(3)
12	(3)
11	(4)
10	(5)
9, 6	Full (0)
5, 5, 5	Full (0)
4, 4, 3, 3	(1)
3, 2, 2, 1	(7)

Figure 7P: The next-fit decreasing algorithm for stacking a bookshelf.

The first-fit-decreasing algorithm tells us to once again use our decreasing list and place the first item on the first shelf. This time we do not close off a shelf. We take the second item and place it on the first shelf it fits and so on. For this example, first-fit-decreasing allows us to completely fill the top six shelves, use part of the seventh shelf, and leave the eighth shelf completely open.

12, 3	Full
12, 3	Full
11, 4	Full
10, 5	Full
9, 6	Full
5, 5, 4, 1	Full
3, 2, 2	(8)

Figure 7Q: The first-fit decreasing algorithm for stacking a bookshelf.

The last algorithm we will discuss is the worst-fit-decreasing algorithm. In this algorithm we place the next item in the first bin with the most space available. The only difference in our results

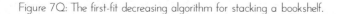

from first-fit-decreasing and worst-fit-decreasing is that the single book would be placed on the seventh shelf where more room is available rather than filling the sixth shelf as we did previously.

12, 3	Full
12, 3	Full
11, 4	Full
10, 5	Full
9, 6	Full
5, 5, 4	(1)
3, 2, 2, 1	(7)

Figure 7R: The worst-fit decreasing algorithm for stacking a bookshelf.

IN- AND OUT-OF-CLASS PROJECTS

1. Get a map of your campus and construct a Hamiltonian route in which to take high school seniors for a tour. You must visit registration offices, financial aid office, bookstore, student union, math offices, English offices, activity center, and the tutoring centers (use as many as your campus offers).

2. Consider both your morning routine and your bedtime routine to construct an order requirement digraph with time estimates and priority list. Determine the length of the critical path and therefore find how long it takes you to get ready in the morning and how long it takes you to get ready for bed each night.

REFERENCES

Investopedia.com—compounding interest

ehow.com—What deductions from a paycheck are reasonable for a worker to expect

betstarter.com—lottery information

horseracing.about.com/cs/handicapping/a/aaoddschart.htm—betting on horseraces

historyoflottery.com—lottery history

VIDEO LIST

CHAPTER ONE

What is compound interest? http://www.investopedia.com/video/play/what-is-compound-interest/
Credit card history. <http://www.pbs.org/wgbh/pages/frontline/shows/credit/>

CHAPTER TWO

Understanding nutrition labels. https://www.youtube.com/watch?v=A2mCfUxhgGc
Calories In vs Calories Out. https://www.youtube.com/watch?v=ETlxpIuXDXw

CHAPTER THREE

What to do if you win the lottery http://www.youtube.com/watch?v=GaBWCBeNl80
How odds work - http://www.youtube.com/watch?v=Jv1n-az457g
Betting at the Track - http://www.youtube.com/watch?v=BpjUCpXPnQg

CHAPTER FOUR

What are statistics http://www.youtube.com/watch?v=p4m9E4f6vwU
Don't be fooled by bad statistics http://www.youtube.com/watch?v=jguYUbcIv8c

CHAPTER FIVE

How barcodes work http://www.youtube.com/watch?v=e6aR1k-ympo

CHAPTER SIX

How to draw in perspective: http://www.youtube.com/watch?v=felys-u4nfk

How to draw a golden rectangle: http://www.youtube.com/watch?v=TLxmLo0ZIg8.

CHAPTER SEVEN

Eulerizations http://www.youtube.com/watch?v=VfHNrZenL0U

Traveling Salesman Problem Visualization http://www.youtube.com/watch?v=SC5CX8drAtU

IMAGE CREDITS

1. Figure 1A. A sample pay check with stub. Source http://www.themint.org/teens/decoding-your-paycheck.html.

2. Figure 2A. Nutrition label serving size. Source: http://commons.wikimedia.org/wiki/File:Nutrition_label.gif. Copyright in the Public Domain.

3. Figure 2B. Nutrition label calorie content. Source: http://commons.wikimedia.org/wiki/File:Nutrition_label.gif. Copyright in the Public Domain.

4. Figure 2C. Nutrition label nutrient content. Source:http://commons.wikimedia.org/wiki/File:Nutrition_label.gif. Copyright in the Public Domain. Source: http://commons.wikimedia.org/wiki/File:Nutrition_label.gif. Copyright in the Public Domain.

5. Figure 2D. Nutrition label nutrient content. Source: http://commons.wikimedia.org/wiki/File:Nutrition_label.gif. Copyright in the Public Domain

6. Figure 2E. Nutrition label percent daily values. Source:http://commons.wikimedia.org/wiki/ File:Nutrition_label.gif. Copyright in the Public Domain

7. Figure 3C. American roulette wheel. Source: http://commons.wikimedia.org/wiki/File:American_roulette_wheel_layout.gif. Copyright in the Public Domain.

8. Figure 3D. American roulette table. Source: http://commons.wikimedia.org/wiki/File:Roulettelayout.svg. Copyright in the Public Domain.

9. Figure 3F. Blackjack reference card. Source: http://www.online-casinos.com/blackjack/blackjack_chart.asp.

10. Figure 4E. Two normal distributions, both with mean 52. Notice the smaller standard deviation is taller and thinner whereas the larger standard deviation is shorter and wider. Copyright © Izabela 09k (CC BY-SA 3.0) at http://commons.wikimedia.org/wiki/File:Assssss.gif.

11. Figure 4K. Generalized multi-bar chart, stacked chart(A), pictograph, and geographical graph (B). Source (A): http://commons.wikimedia.org/

wiki/File:Picture_Graph.svg. Copyright in the Public Domain. Source (B): Copyright © 2013 Depositphotos Inc./Mikko Lemola.

12. Figure 5B. An ISBN 10-digit and 13-digit. Copyright © Littletung (CC BY-SA 3.0) at http://commons.wikimedia.org/wiki/File:ISBN.gif.

13. Figure 5C. UPC label. Copyright © PhilFree (CC BY-SA 3.0) at http://commons.wikimedia.org/wiki/File:Upc.JPG.

14. Figure 5D. Sample check. Copyright © 2012 Depositphotos Inc./John Takai.

15. Figure 5E. Generic credit card. Copyright © 2010 Depositphotos Inc./Laurent Renault.

16. Figure 6A. The first few regular polygons. Source: http://www.shivacharity.com/polygons.html. Copyright in the Public Domain.

17. Figure 6B. Classifications of triangles by side and by angle. Source: http://staff.argyll.epsb.ca/jreed/math7/strand3/3204.htm.

18. Figure 6C. Archimedes' method of inscribing polygons to estimate the value of π. Copyright © KSmrq (CC BY-SA 3.0) at http://commons.wikimedia.org/wiki/File:Archimedes_circle_area_proof_-_inscribed_polygons.png.

19. Figure 6F. The Parthenon of Athens, Greece. Source: http://commons.wikimedia.org/wiki/File:Parthenon.jpg. Copyright in the Public Domain.

20. Figure 6G. The Arc de Triomphe of Paris, France. Copyright © Murali Mohan Gurram (CC BY-SA 3.0) at http://commons.wikimedia.org/wiki/File:ARC_de_TRIUMPHE-PARIS-Dr._Murali_Mohan_Gurram_%282%29.jpg.

21. Figure 6H. The Taj Mahal in India. Copyright © Deep750 (CC BY-SA 3.0) at http://commons.wikimedia.org/wiki/File:Taj_Mahal_in_March_2004.jpg.

22. Figure 6N. (a) Symmetries of a triangle, (b) square, (c) pentagon and (d) hexagon. Sources: (a) Copyright © Fede Threepwood (CC BY-SA 3.0) at http://commons.wikimedia.org/wiki/File:Esferic%C3%B3n_corte_triangular.png. (b) Copyright © Kokcharov (CC BY-SA 3.0) at http://commons.wikimedia.org/wiki/File:LinesOfSymmetryInASquare.png. (c) Copyright © Fede Threepwood (CC BY-SA 3.0) at http://commons.wikimedia.org/wiki/File:Esferic%C3%B3n_corte_pentagonal.png. (d) Source: http://commons.wikimedia.org/wiki/File:Hexagon_Reflections.png. Copyright in the Public Domain.

23. Figure 6O. A tetrahedron. Copyright © Wmheric (CC BY-SA 3.0) at http://commons.wikimedia.org/wiki/File:Schlegeldiagramm_des_Tetraeders.svg.

24. Figure 6P. A cube. Source: http://commons.wikimedia.org/wiki/File:Cubic_graph.svg. Copyright in the Public Domain.

25. Figure 6Q. An octahedron. Copyright © Rob Hooft (CC BY-SA 3.0) at http://commons.wikimedia.org/wiki/File:Achtvlak.png.

26. Figure 6R. A dodecahedron. Source: http://commons.wikimedia.org/wiki/File:Chambers_1908_Dodecahedron.png. Copyright in the Public Domain.

27. Figure 6S. An icosahedron. Copyright © Peter Steinberg (CC BY-SA 3.0) at http://commons.wikimedia.org/wiki/File:Duality_Iko-Dodek.png.

28. Figure 6V. (a) The one point perspective with horizon line and vanishing point. (b) Two point perspective shows both vanishing points. Sources: (a) Copyright © Ejahng (CC BY-SA 3.0) at http://commons.wikimedia.org/wiki/File:Perspective2.jpg. (b) Copyright © Ejahng (CC BY-SA 3.0) at http://commons.wikimedia.org/wiki/File:Perspective1.jpg.

29. Figure 6W. The Last Supper by Leonardo da Vinci Label. Source: http://commons.wikimedia.org/wiki/File:Leonardo_da_Vinci_%281452-1519%29_-_

CPSIA information can be obtained
at www.ICGtesting.com
Printed in the USA
LVOW06s2355080816
499538LV00003B/7/P